Study Guide

ENVIRONMENTAL
SCIENCE
Fifth Edition

Study Guide

Clark E. Adams
TEXAS A & M UNIVERSITY

ENVIRONMENTAL

SCIENCE

Fifth Edition

BERNARD J. NEBEL
CATONSVILLE COMMUNITY COLLEGE

RICHARD T. WRIGHT
GORDON COLLEGE

PRENTICE HALL, UPPER SADDLE RIVER , NJ 07458

Production Editor: *Lisa Protzmann*
Acquisitions Editor: *Sheri L. Snavely*
Supplement Acquisitions Editor: *Mary P. Hornby*
Production Coordinator: *Julia Meehan*

Printed in the United States of America

10 9 8 7 6 5 4 3 2 1

ISBN: 0-13-381419-x

Prentice-Hall International (UK) Limited, *London*
Prentice-Hall of Australia Pty. Limited, *Sydney*
Prentice-Hall Canada Inc., *Toronto*
Prentice-Hall Hispanoamericana, S.A., *Mexico*
Prentice-Hall of India Private Limited, *New Delhi*
Prentice-Hall of Japan, Inc., *Tokyo*
Simon & Schuster Asia Pte. Ltd., *Singapore*
Editora Prentice-Hall do Brasil, Ltda., *Rio de Janeiro*

TABLE OF CONTENTS

TO THE STUDENT

HOW TO USE THIS STUDY GUIDE

I developed this study guide by playing the role of the student who would be using it. For example, you will note that the study guide topics are correlated with and in the same sequence as presented in the text. As you read each chapter of the text, you should easily find the answers to each question. However, this will not be the case if you are using an earlier edition of Environmental Science: The Way the World Works. This study guide was developed specifically for the 5th edition of this text.

Each chapter of the study guide was constructed with three questions in mind, i.e., where are the students heading, how will they get there, and how will they know they have arrived? The introductory statement tells you where you are going by highlighting the particular issues and problems addressed in each chapter. The study questions and vocabulary drill were constructed to give you a directed system of learning thus minimizing wasted effort. Your success on the self-tests should give you some indication of whether you have indeed arrived at one level of mastery of chapter content.

There are two ways you can use this study guide that will help you reinforce text content. One way is to first read an entire chapter and then attempt to answer as many of the study guide and vocabulary drill questions as possible. The correlations of study guide and textbook chapters and subsections should make it easy to look up answers. Any answers you need to look up require another reading of this section of the text and will alert you to content areas that need your additional attention.

Another way to use this study guide is to answer each study quide and vocabulary drill question as you read the text. First read the study guide question, then read the text until you find the answer to that particular question. You will find that using this system will, at times, require that you read several pages before the answer to a particular question is revealed. This system of learning works because it focuses your study on one issue at a time.

After you have attempted to answer all study guide and vocabulary drill questions, check your answers with the answer keys at the end of the study guide. Finally, answer the self-test questions and grade yourself using the self-test answer keys at the end of the study guide.

Remember that incorrect responses to either the study guide, vocabulary drill, or self-test questions are as valuable indicators of your mastery of chapter content as are correct answers. The incorrect responses alert you to specific areas that need additional attention and thus focuses your study time.

Concentrated study of the information in the text and study guide will give you a comprehensive understanding of the pertinent ecological principles, how various human activities are in conflict with these principles, and how this conflict might be mitigated by redirecting human activities to be in concert with ecological principles. If you can address each issue on these three levels, then you should enjoy a high level of personal satisfaction and academic success.

CHAPTER 1

INTRODUCTION: ENVIRONMENTAL SCIENCE AND SUSTAINABILITY

Whether one lives in Tanzania, the Amazon Jungle, or Harlem or Hollywood in the United States, all human life depends on a stable global environment including the ecological, cultural, economic, and political components. We need to rely on our environment to provide the basic human needs of nutritious food, clean water, a safe place to live, an education, health care, and a job. Once the basic needs have been met, we humans, unlike other species, need to accumulate possessions (e.g., clothing, cars, tools, and other STUFF). The provision of basic needs and possessions causes many environmental problems because the activities involved in resource use, e.g., access, withdrawal, production, transportation, distribution, and waste management, are not linked to a human philosophy of sustainability. Furthermore, since equitable distribution of resources is not reality in human (or natural) ecosystems, there will always be a societal division of "haves and have nots," real or imagined. Is the concept of "sustainability" an achievable goal given the paradox of limited resources and an ever increasing human population with accompanying basic and possession needs? The answer is yes if humankind adopts and adapts to new models of co-existence with the natural world. This chapter introduces you to the essence of the new models and how they can be achieved. We are at a point in the history of human presence on this earth that the development and implementation of these new models is an obligation and opportunity.

Are you a cornucopian or an environmentalist? If your world view is to spend the interest and capital of your natural resource inheritance, then you are a cornucopian. Get ready for some hard times since your world view will be challenged on every front as you progress through this course of study. Conversely, if your world view is to spend only the interest from your natural resource inheritance, then you are an environmentalist. Your world view will be supported and nurtured in this class. You will be given the tools of logic and knowledge to combat the cornucopian world view.

STUDY QUESTIONS

1. The historic residents of Easter Island learned that civilization collapses when:

   a. _____

   b. _____

   c. _____

2. People who pay close attention to connections between people and their environment are called

   _____

Environmentalism

3. Trace the conservation attitude of Americans during the following time periods and by indicating whether the attitude was positive [+] or negative [-].

    a. [   ] late nineteenth century
    b. [   ] scientific era after World War I
    c. [   ] depression years (1930 - 1936)
    d. [   ] during World War II
    e. [   ] after World War II until the 1960s
    f. [   ] from the mid 1960s to 1980
    g. [   ] 1980 to now

4. What appeared to be the two most important factors that stimulated a conservation mood compared to periods when this mood was not evident?

    a. _____ b. _____

Silent Spring, Loud Reaction

5. Identify one environmental problem that developed during the post-World War II era in the

    a. Air: _____

    b. Waters: _____

    c. Wildlife: _____

6. What was industry producing that caused such dramatic pollution problems?

    _____

7. What "cause and effect" relationship did Rachel Carson portray in her book titled: Silent Spring?

    _____

8. Why does the position of citizens aligned with the **environmental movement** include wildlife?

    _____

    _____

9. Identify four environmental organizations that grew out of the environmental movement.

    a. _____ b. _____

    c. _____ d. _____

Early Results Were Encouraging

10.   List four successes associated with the environmental movement.

      a. _____

      b. _____

      c. _____

      d. _____

11.   List three vocations (i.e., "environmental") that grew out of the environmental movement.

      a. _____   b. _____

      c. _____

Environmentalism Acquires Its Critics

12.   Give an example of the following sources of pollution.

      a. point: _____

      b. diffuse: _____

13.   What are the two major points of contention in most environmental conflicts?

      a. _____ vs. b. _____

14.   What new movement seems to be the antitheses of the environmental movement?

      _____

The Global Environmental Picture

15.   What world-wide events indicate that environmentalism is "losing the war"?

      a. _____   b. _____

      c. _____   d. _____

Population Growth

16.   Give the world population figures in:

      a. 1995 _____   b. 2050 _____

17. Given the rapid human population growth rate and expected global population size by 2050, discuss the growing conflict between "global quality of life" and "American lifestyle".

_____

_____

Degradation of Soils

18. What are the indicators of global soil degradation?

    a. _____   b. _____

    c. _____   d. _____

    e. _____

Atmospheric Changes

19. Give one example of an air pollution problem with global dimensions.

    _____

20. The carbon dioxide level in the atmosphere has grown from _____ parts per

    million (ppm) in 1900 to _____ ppm today.

21. Explain the expected relationship between atmospheric carbon dioxide and global warming.

_____

_____

Loss of Biodiversity

22. List three ways in which **habitat alteration** is taking place.

    a. _____   b. _____

    c. _____

23. What is meant by the term **biodiversity**?

_____

24. What three factors are the greatest contributors to loss of biodiversity?

    a. _____   b. _____

    c. _____

25. Give three reasons why the loss of biodiversity is so critical.

    a. _____

    b. _____

    c. _____

Toward the Future

26. Give two predictable outcomes resulting from a continuance of global trends that are *not sustainable*.

    a. _____

    b. _____

27. What was the prediction of Thomas Malthus in 1798 concerning food production and human population growth?

    _____

    _____

28. What was wrong with the Malthus prediction at the time he made it?

    _____

    _____

29. List two contemporary trends that support the Malthus prediction into the 21st century.

    a. _____

    b. _____

30. The world population is now growing _____ times faster than in Mathus's time.

31. Indicate whether we can [+] or cannot [-] rely on the following technologies to produce enough food for the world population in the future.

    [　] irrigation of dry land
    [　] chemical fertilizers
    [　] new genetic varieties of crop plants and animals
    [　] bringing more agricultural land into production

32.   Define the term **cornucopian**.

_____

_____

33.   Indicate whether the following assumptions or statements characterize the cornucopian [C] or environmentalist [E] world view.

[   ] Natural resources are limited.
[   ] All natural resources are to be exploited for the advantage of humans.
[   ] Maintain the status quo on human uses of natural resources.
[   ] Change the direction of human uses of natural resources.
[   ] It represents a sustainable philosophy.
[   ] It is the dominant world view.

Sutainable Development

34.   Define the term **sustainable**.

_____

_____

35.   The current trends of human use of the world's natural resources (Is, Is Not) sustainable.

36.   Imagine you are in the commercial fishing (e.g., tuna) business. Fisheries biologists have estimated that the fishery consists of 10K tuna of which 8K are the surplus resulting from the reproducing component of the population. In order to maintain a **sustainable yield** how many tuna should be harvested? _____

37.   Justify your answer to question #36.

_____

_____

38.   List three characteristics of a **sustainable society**.

a. _____

b. _____

c. _____

A New Meaning of Development

39.   Define the term **sustainable development**.

_____

40.   Apply the meaning of sustainable development to:

    a. soil: _____

    b. wildlife: _____

    c. population: _____

    d. crime: _____

    e. race: _____

    f. education: _____

    g. technology: _____

The New Environmentalist

41.   List three characteristics of the contemporary environmentalist.

    a. _____

    b. _____

    c. _____

VOCABULARY DRILL

Directions: Match each term with the list of definitions and examples given in the tables below.

Term	Definition	Example	Term	Definition	Example
biodiversity			habitat alteration		
cornucopian			sustainable yields		
environmental vocations			sustainable society		
environmental movement			sustainable development		
environmentalist			sustainable		

VOCABULARY DEFINITIONS
a. people who pay close attention to the connections between people and their environment
b. militant citizenry demanding curtailment and cleanup of pollution and protection of pristine environments
c. habitat changes that may or may not be reversible
d. the total assemblage of all plants, animals, microbes that inhabit the earth
e. attitude that new technologies will fix all problems
f. continued indefinitely without depleting any of the material or energy resources to keep it running
g. harvest of renewable resources that does not exceed replacement rates
h. continues without depleting its resource base or producing pollutants
i. progress that meets the needs of the present without compromising the needs of future generations
j. job opportunities that require environmental science backgrounds

VOCABULARY EXAMPLES
a. Rachel Carson
b. Earth Day 1970
c. Clear-cut logging
d. Global flora and fauna
e. Any one who does not think like an environmentalist
f. Infinity
g. American indians
h. *Our Common Future*
i. Economics, engineering, scientist
j. Cashing in on the interest only

SELF-TEST

1. Rank (1 = first event to 5 = last event) the following chain of events in the order they would occur leading to the collapse of a civilization.

Events   Rank

a. loss of topsoil   [   ]
b. population increase beyond food growing capacity   [   ]
c. deforestation   [   ]
d. soil erosion   [   ]
e. disparity between haves and have nots   [   ]

2.   Explain why the Easter Island population could have been more susceptible to over population and resource depletion than continental populations?

_____

_____

3.   During which of the following time periods did Americans have a positive attitude toward conservation?

    a. scientific era after World War I   b. depression years
    c. during World War II   d. after World War II until the 1960s

4.   In general, a positive conservation attitude by the majority of Americans is initially stimulated by:

    a. public education   b. environmental crisis
    c. love of nature   d. federal laws

5.   Which of the following was not a post-World War II environmental problem in America?

    a. air pollution   b. decline of wildlife populations
    c. water pollution   d. over population

6.   Indicate whether the following are gains [+] or losses [-] in the war for the environment.

    a. [   ] population growth   b. [   ] pollution control
    c. [   ] environmental laws   d. [   ] biodiversity

7.   Check the most important initial battle to win in the war for the environment.

    a. [   ] population growth   b. [   ] pollution control
    c. [   ] environmental laws   d. [   ] biodiversity

8.   During the spotted owl controversy in the Northwest, the debate was over the
_____ versus _____.

9.   Match the evidence with list of global events indicating that environmentalism is losing the war.

Evidence

[   ] habitat conversions
[   ] CO2 emissions
[   ] desertification
[   ] 10 billion
[   ] commercial exploitation
[   ] urban development

Events

1. population growth
2. atmospheric changes
3. soil degradation
4. loss of biodiversity

10. Given a continuance of present global trends, apply the predictions of Thomas Malthus to the year 2050.

   _____

   _____

11. Are you a cornucopian or an environmentalist? Below is a list of world view assumptions. Check all those that apply to your personal world view regarding human uses of natural resources.

   [   ]   Natural resources are to be exploited for the advantage of humans.
   [   ]   Natural resources are essentially infinite.
   [   ]   We need to stay on the same course of human uses of natural resources that has been the historical trend of Western civilization.
   [   ]   It is alright to spend both the interest and the capital of our natural resource inheritance.
   [   ]   Natural resources are to be protected and maintained.
   [   ]   Natural resources are limited by the regenerative capacities of the natural environment.
   [   ]   We need to change the ways humans use natural resources.
   [   ]   We need to live only on the interest of our natural resource inheritance.

12. Any actions that promote a sustainable human use of natural resources will require public

   a. understanding and awareness of the problems.
   b. adoption of new values.
   c. demands for change.
   d. all of the above

True [+] of False [-]

13.   [   ] American's attitudes toward conservation has been consistent throughout the 20th century.

14.   [   ] Environmentalists pretty much dictate the future use of global resources.

15.   [   ] Air and water in America is cleaner now than in the 1960s.

16.   [   ] The outcomes of global trends that are not sustainable are predictable.

17.   [   ] The outcomes of establishing a sustainable society are predictable.

18.   [   ] The Wise Use and Environmental Movements have compatible goals.

19.   [   ] The present American lifestyle is compatible with a sustainable society.

20.   [   ] The characteristics of the contemporary environmentalists pretty much describe the majority of Americans.

CHAPTER 2

ECOSYSTEMS: UNITS OF STABILITY

This chapter addresses the complexity of ecosystem structure as units of stability that can be explained and interpreted. Ecosystem complexity is simplified by understanding the overriding blueprint of ecosystem structure. First, there are fundamental building blocks, e.g., abiotic, plant, and animal components. Second, there is a predetermined consistent order for putting the building blocks together, i.e., the abiotic components predict the plant component which predicts the animal component. Third, predictions can be made concerning outcomes from changes in the building blocks or the order for putting them together. For example, any change in the abiotic conditions will result in changes in the plant and consequently animal components. Furthermore, the introduction of plants or animals (exotics) into abiotic conditions that did not initially predict their presence usually leads to disastrous consequences for the exotics and/or endemic plants and animals. Ecosystems are units of stability because of the above blueprint.

The complexity of ecosystem structure is also examined in terms of an ecological conceptual framework that will apply to nearly every chapter in this text. This conceptual framework consists of four parts including cycles, energy, diversity, and interrelationships. Consistent application of this framework to your study of ecology and how the world works will give you the principle focal points for analysis of chapter content.

Chapter 2 provides opportunity to study various *natural ecosystems* in terms of the basic parts and how they fit together to make a working system. There are many different kinds of ecosystems but the fundamental parts and the way they work is the same. Thus the terminology and concepts presented will give a basic foundation for discussing and understanding all ecosystems and the human impacts on each.

STUDY QUESTIONS

What Are Ecosystems?

1. All **biotic communities** consist of _____, _____, and _____ communities.

2. Identify the following community components as **abiotic** [A] or **biotic** [B].

    a. [   ] plants   b. [   ] animals   c. [   ] temperature

    d. [   ] microbes   e. [   ] rainfall   f. [   ] wind

3. An organism can be designated a distinct **species** if it can _____ with others of its own kind and produce _____ offspring.

4. A group of bullfrogs (<u>Rana</u> <u>catesbiana</u>) is the same farm pond constitutes a _____.

5. An **ecosystem** may be defined as a grouping of _____, _____, and _____ interacting with _____ and their _____.

6. Furthermore, the interrelationships are such that the grouping may be _____.

7. List six examples of major kinds of ecosystems.

   a. _____   b. _____   c. _____

   d. _____   e. _____   f. _____

8. Each ecosystem is characterized by distinctive plant _____.

9. Identify the **ecotone** in Fig. 2-2. _____

10. The major **biome** of eastern United States is _____.

11. The major biome of central United States is _____.

12. The major biome of southwestern United States is _____.

13. The distribution of major biomes globally is the consequences of _____.

14. Is it (True or False) that what occurs in one ecosystem or biome may affect other ecosystems or biomes?

15. Give two ways that terrestrial ecosystems remain connected even though they may be far apart (.e.g., coniferous forests in the northwest with southwest grasslands).

   a. _____

   b. _____

16. All creatures on earth and the environments they live in constitute one huge ecosystem called the _____.

The Structure of Ecosystems

17. The two basic structural components of all ecosystems are the _____ communities and _____ environmental factors.

Biotic Structure

Categories of Organisms

18. All the organisms in an ecosystem can be assigned to one of three categories. These three categories are _____, _____, and _____.

19. Producers include all plants that do what? _____

20. In photosynthesis, plants use _____ energy to convert _____ gas and _____ into _____ .

21. The word **inorganic** refers to substances such as _____ .

22. The chemicals that make up the bodies of organisms are referred to as _____ .

23. In short, producers convert (Inorganic, Organic) chemicals into (Inorganic, Organic) chemicals.

24. The energy for this production comes from _____ .

25. Organisms which can produce all their own organic materials from inorganic raw materials in the environment are called (Autotrophs, Heterotrophs). Give an example _____

26. Organisms which must consume organic material as a source of nutrients and energy are called (Autotrophs, Heterotrophs). Give an example _____

27. Is it (True or False) that **all** plants are autotrophs? If you said false, give an example of an exception. _____

28. List three organisms that could be classified as consumers.

    a. _____   b. _____   c. _____

29. Consumers that feed directly on producers are called **primary consumers** or _____ .

30. List two organisms that are primary consumers.

    a. _____   b. _____

31. Consumers that feed on other consumers are called secondary consumers or _____ or _____ .

32. Indicate whether the following list of animals would be **predators** [PR] or **prey** [PY] in **predator/prey** relationships.

    a. [   ] wolf   b. [   ] coyote   c. [   ] quail
    d. [   ] bear   e. [   ] deer   f. [   ] salmon

33. Give an example of:

    a. internal parasite: _____   b. external parasite: _____

    c. parasite host: _____

34. All dead organic matter (dead leaves, twigs, bodies of animals, fecal matter) is called _____ .

35.   Give three examples of **detritus feeders.**

a. _____ , b. _____ , c. _____

36.   **Decomposers** consist of _____ and _____ .

Feeding Relationships: Food Chains, Food Webs, and Trophic Levels

37.   Construct a **food chain** using a grassland community that consists of [a] big bluestem grass, [b] grasshoppers, [c] frogs, [d] snakes, and [e] owls. First consider what each organism eats. For example:

a. grasshoppers eat: _____   b. frogs eat: _____

c. snakes eat: _____   d. owls eat: _____

e. Next, place each animal in a sequence of **feeding relationships** [a to e] that demonstrates a food chain.

[   ] ⇨ [   ] ⇨ [   ] ⇨ [   ] ⇨ [   ]

38.   Let's add some more organisms to the above grassland community including: fescue grass, berry bushes, mice, white-tailed deer, humans, rabbits, foxes, red-tailed hawk, ground squirrels, and sunflowers. Construct a **food web** in the below diagram by placing each species in its proper **trophic level** and then drawing lines of feeding relationships (see Fig. 2-13a for an example of this exercise). The species to use in this exercise are as follows: a. bluestem grass, b.grasshopper, c. frogs, d. snakes, e. owls, f. fescue grass, g. berry bushes, h. mice, i. deer, j. humans, k. rabbits, l. foxes, m. hawks, n. ground squirrels, and o. sunflowers

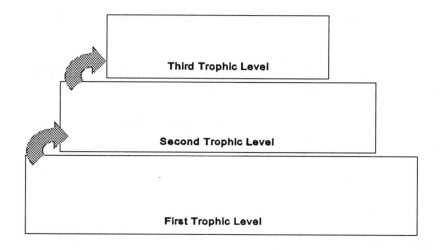

39. Indicate whether the following categories of organisms occupy the [a] first, [b] second, [c] third, or [d] more than one trophic level.

a. [   ] producers   b. [   ] consumers   c. [   ] decomposers

d. [   ] carnivores   e. [   ] herbivores   f. [   ] predator

g. [   ] prey   h. [   ] omnivores   I. [   ] parasites

j. [   ] autotrophs   k. [   ] heterotrophs

40. How is **biomass** determined?

_____

_____

41. In an ecosystem the biomass of herbivores is always (Greater, Less) than the biomass of producers, and the biomass of carnivores is always (Greater, Less) than the biomass of herbivores.

42. Diagram the biomass relationship for three trophic levels.

Nonfeeding Relationships. Mutually Supportive Relationships

43. Describe an example of **mutualism**.

_____

44. How does your example of mutualism demonstrate the concept of **symbiosis**?

_____

Competitive Relationships

45. Where an organism lives is called its (habitat or niche). How an organism lives is called its (habitat or niche).

46. Give an example of how interspecies niche competition is reduced in:

a. space: _____

_____

b. time: _____

Abiotic Factors

47.   Three examples of abiotic factors are:

   a. _____ b. _____ c. _____

Optimum Zones Of Stress, and Limits of Tolerance

48.   Label the below figure with the following terms (see Fig. 2-19 for an example of this exercise). **a. optimum, b. range of tolerance, c. limits of tolerance, d. zones of stress**

49.   As the amount or level of any factor increases or decreases from the optimum, the plant or animal is increasingly _____.

50.   If any factor increases or decreases beyond an organism's limit of tolerance, what occurs? _____.

51.   Is it (True or False) that each species has an optimum zone of stress, and limits of tolerance with respect to every factor?

Law of Limiting Factors

52.   How many **limiting factors** need to be beyond a plant's limit of tolerance to preclude its growth? _____

53.   Give two examples of limiting factors.

   a. _____   b. _____

Why Do Different Regions Support Different Ecosystems?

Climate

54.   The definition of **climate** involves both _____ and _____.

55.   _____ is the main climatic factor responsible for the separation of ecosystems into forests, grasslands, and deserts.

56.   Even though it takes 30 inches or more of rainfall to support a forest ecosystem, _____ will determine the kind of forest.

57.   Give an example of an ecosystem that would contain **permafrost**. _____

Microclimate and Other Abiotic Factors

58.   A localized area within an ecosystem that has moisture and temperature conditions different from the overall climate of the area is called a _____.

Biotic Factors

59.   The biotic factor that limits grass from taking over high rainfall regions is

        a. tall trees   b. herbivores   c. parasites   d. insects

Physical Barriers

60.   The spread of species into other ecosystems may be prevented by natural physical barriers such as _____, _____, and _____.

61.   The spread of species into other ecosystems may be prevented by physical barriers produced by humans such as _____, _____, and _____.

Implications for Humans

62.   As hunter-gatherers, humans were much like: a. herbivores, b. carnivores, c. omnivores

63.   With _____, humans created their own distinctive ecosystem apart from natural ecosystems.

64.   List five ways humans have overcome the usual limiting factors that prevent the spread of our species.

        a. _____

        b. _____

        c. _____

        d. _____

        e. _____

65.   With the proliferation and spread of the **human system** in its present context, is it (True or False) that we are upsetting and destroying other ecosystems?

			VOCABULARY DRILL		
Directions:			Match each term with the list of definitions and examples given in the tables below. Term definitions and examples may be used more than once.		
Term	**Definition**	**Example (s)**	**Term**	**Definition**	**Example (s)**
abiotic			heterotroph		
abiotic factors			host		
autotroph			inorganic molecule		
biomass pyramid			limiting factors		
biomass			limits of tolerance		
biome			microclimate		
biosphere			mutualism		
biota			niche		
biotic community			omnivore		
biotic factors			optimum		
biotic structure			organic molecule		
carnivore			parasites		
chlorophyll			parasitism		
climate			photosynthesis		
consumers			population		
decomposer			predator		
detritus			prey		
detritus feeders			primary consumer		
ecologist			producers		
ecology			range of tolerance		
ecosystem			secondary consumer		
ecotone			species		
food chain			symbiotic		
food web			synergistic effect		
habitat			trophic levels		
herbivore			zone of stress		

VOCABULARY DEFINITIONS

a. grouping or assemblage of living organisms in an ecosystem

b. assemblage of nonliving physical or chemical components in an ecosystem

c. specific kinds of plants, animals, or microbes that interbreed and produce fertile offspring

d. number of individuals that make up an interbreeding, reproducing group within a given area

e. a grouping of plants, animals, and microbes interacting with each other and their environment in such a way to perpetuate the grouping

f. study of ecosystems and interactions

g. people who study ecosystems and interactions

h. the transitional area where one ecosystem blends into another

i. total combined dry weight of all organisms at a trophic level

j. all the species and ecosystems combined

k. the way different categories of organisms fit together in an ecosystem

l. organisms that produce their own organic material from inorganic constituents

m. organisms that must feed on complex organic material to obtain energy and nutrients

n. the process of converting sunlight energy and CO_2 to sugar and O_2

o. molecule that plants use to capture light energy for photosynthesis

p. constructed in large part from carbon and hydrogen atoms

q. constructed in large part from elements other than carbon and hydrogen

r. dead or partially digested plant or animal material

s. primary detritus feeders

t. feed directly on producers

u. feed on primary consumers

v. meat-eating secondary consumer

w. plant and meat-eating consumer

x. animal that attacks, kills, and feeds on another animal

y. animal killed and eaten by a predator

z. a predator that feeds off its prey for a long time typically without killing it

aa. plant or animal fed upon by a parasite

bb. pathways where one organism is eaten by a second, which is eaten by a third, and so on

VOCABULARY DEFINITIONS
cc. a way of demonstrating the interrelatedness of food chains
dd. feeding levels within a food chain or web
ee. shape of biomass potential at each trophic level
ff. a symbiotic relationship between two organisms in which both derive benefit
gg. living together
hh. a symbiotic relationship between organisms in which one benefits and the other is harmed
ii. plant community and physical environment where an organism lives
jj. what an organism feeds on, where and when it feeds, where it finds shelter and nesting sites
kk. physical and chemical components of ecosystems
ll. a certain abiotic level where organisms survive best
mm. abiotic level between optimal range and high and low level of tolerance
nn. high and low ranges of tolerance to abiotic levels
oo. any factor that limits the growth, reproduction, and survival of organisms
pp. greater effect of two factors interacting together than individually
qq. description of average temperature and precipitation each day throughout the year
rr. conditions in specific localized areas
ss. limiting factors caused by other species
tt. grouping of related ecosystems into major kinds of ecosystems
uu. the entire span of abiotic values that allow any growth at all
vv. organisms that feed primarily on dead, decaying, or partially digested organic matter

VOCABULARY EXAMPLES	
a. plants, animals, microbes	r. dead leaves
b. CO2, rainfall, pH, O2, salinity, temperature	s. bacteria or fungi
c. Rana catesbiana (bullfrogs)	t. deer or rabbit
d. All Rana catesbiana in a farm pond	u. wolf or owl
e. Fig 2-2	v. pigs and bears
f. what you are doing now	w. Fig 2-9
g. your teacher	x. Fig 2-13a
h. where grassland meet forest	y. Figs. 2-16 and 2-17
i. Fig. 2-14	z. Fig. 2.8 or Fig. 2.9
j. Earth	aa. Fig. 2-19
k. producers, consumers, decomposers	bb. chemical A → effect B, chemical C → effect D, chemical A + B → effect E
l. green plants	cc. Fig. 2-20
m. herbivores, carnivores, omnivores	dd. Fig. 2-22
n. $6CO_2 + 6H_2O \rightarrow C_6H_{12}O_6 + 6O_2$	ee. Fig. 2-4
o. green organelle in plant tissue	
p. glucose ($C_6H_{12}O_6$)	
q. salt (NaCl)	

SELF -TEST

An ecosystem is best defined as a grouping of 1. _____ ,

2. _____ and 3. _____ interacting with 4. _____

and their 5. _____ in such a way that the grouping is perpetuated.

Ecosystems evolve in sequential patterns involving plant, animal, and abiotic components. What is the component sequence in ecosystem development?

6. First Component: _____   7. Second Component: _____

8. Third Component: _____

9.   A major biome in the central United States is a

      a. deciduous forest, b. grassland, c. desert, d. coniferous forest

10.   Which of the following is not correctly matched?

      a. oak tree - producer
      b. squirrel - consumer
      c. mushroom - detritus feeder
      d. fungi - decomposer

11.   The three major biotic components of ecosystem structure are

      a. producers, herbivores, and carnivores
      b. producers, consumers, and decomposers
      c. plants, animals, and climate
      d. consumers, detritus feeders, and decomposers

      Below is a typical food chain illustration. Answer the series of questions below the illustration by choosing level 1, 2, 3, 4, 5, or ALL.

                              Level 5: OWL
                              Level 4: SNAKE
                              Level 3: FROG
                              Level 2: GRASSHOPPER
                              Level 1: GRASS

At which level:

12. _____ is the most transferable energy available?

13. _____ would one find the least biomass?

14. _____ is a primary consumer located?

15. _____ is a producer located?

16. _____ could a white-tailed deer be a representative?

17. _____ is sunlight the original energy source?

18. _____ would one find decomposers?

19. _____ would one find parasites?

20.   The biomass relationship of a food chain can only be depicted as a (a. box, b. rectangle, c. circle, d. pyramid)

21.   Explain your answer choice to #20.

_____

22. Which of the following is not an example of an abiotic factor?

   a. water, b. light, c. dead log, d. temperature

23. The points at which a factor becomes so high or low as to threaten an organism's survival is referred to as

   a. range of optimums, b. range of limits, c. range of tolerance, d. limits of tolerance

24. How many factors need to be beyond an organism's limits of tolerance to cause stress?

   a. one, b. five, c. ten, d. hundreds

   Which limiting factor (a. temperature, b. rainfall, c. pH, d. soil type) is the best determinant of:

25. _____ deserts, grasslands, and forests?

26. _____ kind of forest?

27. _____ kind of desert?

   Identify those conditions that do [+] and do not [-] represent the human system and then indicate whether the listed conditions are [+] or are not [-] compatible with natural ecosystems.

Human System Conditions	Compatibility with natural ecosystems
28. [ ] Recycle wastes	32. [ ]
29. [ ] Practice population control	33. [ ]
30. [ ] Control limiting factors	34. [ ]
31. [ ] Causing rapid abiotic/biotic changes	35. [ ]

CHAPTER 3

ECOSYSTEMS: HOW THEY WORK

This chapter expands on the previous one by providing a more detailed explanation of interrelationships, cycles, energy, and ecosystem structure and function. Ecosystem structure is explained in terms of identifying those atoms (N, CHOPS) most used to build the living (biotic) component of ecosystems. Ecosystem function is discussed in terms of how producers and consumers use energy and are involved in nutrients cycling. Chapter content may be difficult because there is no personal recognition of things like atoms, molecules, matter, and energy in your day-to-day life. Additionally, students have a hard time grasping how energy is brought into the ecosystem, how it is used, stored, and lost. Ecosystems work because matter (N, CHOPS) is recycled through processes driven by a never-ending supply of solar energy. However, ecosystems use matter and energy differently. Energy use is a one-way trip! It comes into the ecosystem as light energy, captured through photosynthesis, and stored in green plants as glucose. Both plants and animals rely on this initial supply of glucose for all life support functions and activities. Both use cellular respiration to release the energy stored in glucose. All energy is ultimately lost from the ecosystem in the form of heat. If you can grasp the principles of recycling matter, energy flow, and biomass, the concepts presented in this and succeeding chapters will make sense. In fact, you will (should) be shocked by the human implications regarding these concepts.

STUDY QUESTIONS

Elements, Life, Organization, and Energy

1. **Matter**, in a chemical sense, refers to all _____, _____, and _____.

2. **Atoms** or **elements** are the basic building blocks of all _____.

3. How many different kinds of elements occur naturally? _____

4. Only _____ of all the known elements are used to produce living tissues (see Table 3-1, iodine is missing).

5. Indicate whether the following statements are true [+] or false [-] concerning the **Law of Conservation of Matter**.

    [   ] atoms cannot be created
    [   ] atoms can be destroyed
    [   ] atoms can be converted into others
    [   ] atoms can be rearranged to form different molecules

6.   Is it (True or False) that a **molecule** = a **compound**? Is it (True or False) that a compound = a molecule?

7.   Indicate whether each of the following is an element [E] or compound [C].

a. [   ] oxygen b. [   ] carbon dioxide   c. [   ] water   d. [   ] nitrogen
e. [   ] ozone   f. [   ] phosphorus   g. [   ] carbon   h. [   ] protein
i. [   ] sugar   j. [   ] hydrogen

8.   For each of the following elements, which are critical in living organisms, give a compound or molecule which contains it and where it is found in the abiotic environment. (see Table 3-1)

Element	Compound Containing It	Environmental Location
carbon	_____	_____
hydrogen	_____	_____
oxygen	_____	_____
nitrogen	_____	_____
phosphorus	_____	_____

9.   Match the list of elements and molecules.

Elements   Molecules

a. N. _____   1. glucose
b. C. _____   2. proteins
c. H. _____   3. starch
d. O. _____   4. fats
e. P. _____   5. nucleic acids
f. S. _____   6. all of the above

10.   Is it (True or False) that all **organic** compounds are found in living systems.

Energy Considerations

11.   **Matter** is defined as anything that occupies _____ and has

_____ .

12.   From the most to least dense states, matter can exist in the forms of _____,

_____ and _____ .

13.   **Energy** is defined as: _____

14.   List three forms of energy. _____, _____, and

_____

15. Indicate whether the following statements about energy are true [+] or false [-].

    [   ] Energy occupies space.
    [   ] Energy can be weighed.
    [   ] Kinetic energy can be transformed into potential energy.
    [   ] Potential energy can be transformed into kinetic energy.
    [   ] Energy can be recycled.
    [   ] Energy can be created.
    [   ] Energy can be converted from one form to another.
    [   ] There is some loss in each energy conversion.
    [   ] Energy can be measured.

16. Heat, light, and motion are forms of (Kinetic, Potential) energy.

17. Glucose, a compressed spring, and fossil fuels are forms of (Kinetic, Potential) energy.

18. The high potential energy in fuels is commonly referred to as _____ energy.

19. A _____ is the unit measure of the amount of heat required to raise the temperature of 1 gram of water 1 degree Celsius.

Energy Laws: Laws of Thermodynamics

20. Indicate whether the following statements pertain to the [a] **First** or [b] **Second Laws of Thermodynamics**.

    [   ] Energy cannot be created or destroyed.
    [   ] Energy can be converted from one form to another.
    [   ] Energy conversions always result in net losses of useful energy.
    [   ] Heat is the final energy conversion form.

Matter and Energy Changes in Organisms and Ecosystems

Energy and Organic Matter

21. Identify whether the level of potential energy in the following types of molecules is high [H] or low [L].

    [   ] inorganic   [   ] organic   [   ] coal
    [   ] water   [   ] sugar   [   ] wood

Producers

22. In most ecosystems, the primary producers are plants that carry on _____.

23. Light energy for photosynthesis is absorbed by the _____ molecule.

24. The efficiency of photosynthesis in converting light energy to chemical energy is at best _____ percent efficient.

25. The primary product of photosynthesis is _____.

26. List three ways plants use the glucose produced during photosynthesis.

    a. _____

    b. _____

    c. _____

Consumers and Other Heterotrophs

27. The basic source of energy and raw materials for animal growth and developement is
    _____

28. **Cell respiration** is effectively the _____ process of photosynthesis in terms of what is consumed and released.

29. Indicate whether each of the following refers to photosynthesis [P], cell respiration [R], both [B], or neither [N].

    [ ] releases oxygen
    [ ] stores energy
    [ ] releases carbon dioxide
    [ ] consumes carbon dioxide
    [ ] releases energy
    [ ] produces sugar
    [ ] consumes sugar
    [ ] consumes oxygen

30. Place an [x] by any of the following organisms that carry on cell respiration for at least part of their energy needs.

    [ ] consumers   [ ] decomposers   [ ] plants   [ ] fungi
    [ ] bacteria

Nutritive Role of Food

31. The lack of certain nutrients in your diet may lead to various diseases caused by
    _____.

32. The point of a balanced diet is that it supplies both _____ and
    _____ in adequate but not excessive amounts.

Material Consumed, but Not Digested

33. Trace the three pathways of food eaten by any consumer.

    a. _____

b. _____

c. _____

34.   Is it (True or False) that fecal waste is no longer a source of nutrients and energy?

Detritus Feeders and Decomposers

35.   Give two examples of detritus feeders that require a mutualistic relationship with other organims to digest **cellulose**.

   a. _____   b. _____

Principles of Ecosystem Function

Nutrient Cycling

36.   *For sustainability, ecosystems dispose of wastes and replenish nutrients by recycling all elements* is the (First, Second, Third) basic principle of ecosystem sustainability.

37.   Natural ecosystems do not harm themselves with their own waste products because these products are _____.

The Carbon Cycle

38.   The reservoir of carbon for the carbon cycle is _____ present in the
   _____.

39.   Carbon dioxide is first incorporated into sugar by the process of _____ and then it may be passed to other organic compounds in other organisms through food chains.

40.   At any point on the food chain an organism may use the carbon compounds in cell respiration to meet its energy needs. When this occurs, the carbon atoms are released as
   _____.

41.   The amount of atomspheric CO2 is being increased greatly by the burning of
   _____

42.   Complete the diagram of the CARBON CYCLE by putting in all arrows of interrelationships between organisms and labeling the empty boxes (see Fig 3-14 for an example of this exercise).

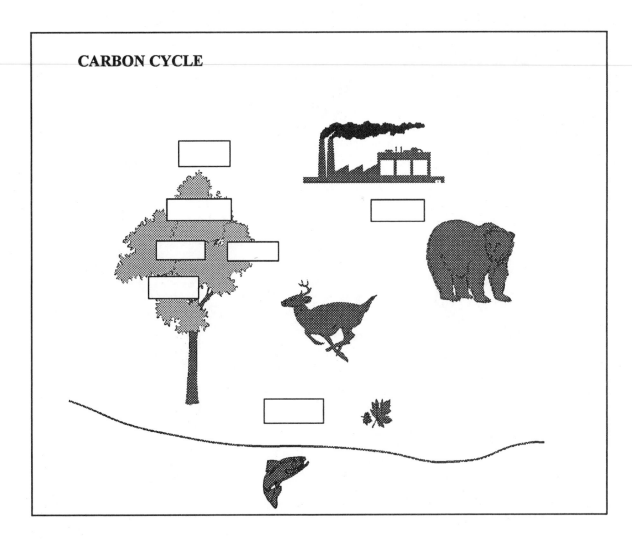

CARBON CYCLE

The Phosphorus Cycle

43. The reservoir of the element phosphorus exists in various rock and soil _minerals_.

44. Plants absorb phosphate from the _soil_ or _water_ solution.

45. Phosphate is first assimilated into organic compounds as _organic_ phosphate and then passed to various other organisms through the food chain.

46. At any point on the food chain, an organism may use organic phosphorus in respiration. When this occurs, the phosphate is released back to the environment in _urine & other wastes_

47. Humans disrupt the phosphorus cycle by _cutting down rain forests_ and _phosphate fertilizer. runoff from agricultural lands._

48. Complete the diagram of the PHOSPHORUS CYCLE by putting in all arrows of interrelationships between organisms and labeling the empty boxes (see Fig 3-15 for an example of this exercise).

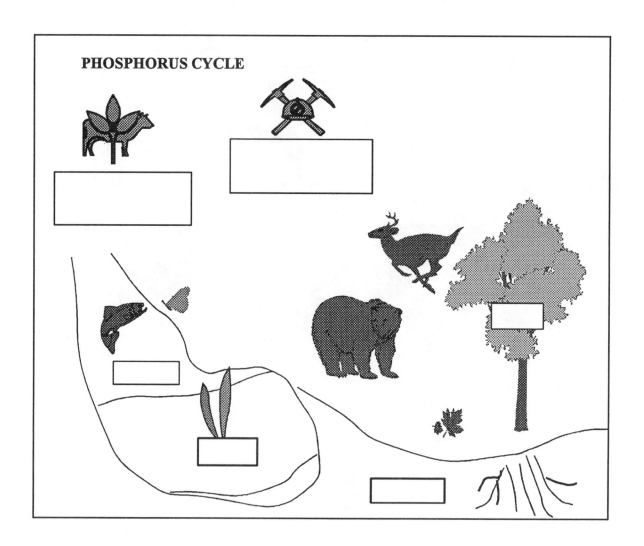

PHOSPHORUS CYCLE

The Nitrogen Cycle

49. The main reservoir of nitrogen is N_2 in the air which is about _____ percent nitrogen gas.

50. Is it (True, False) that plants cannot assimilate nitrogen gas from the atmosphere?

51. To be assimilated by higher plants, nitrogen must be present as _____ or _____ ions.

52. A number of bacteria and certain blue green algae can convert nitrogen gas to the ammonium form, a process called nitrogen _____.

53. The bacteria have a mutualistic relationship with the _____ family of plants.

54. Once fixed, nitrogen may be passed down food chains, excreted as _____ and recycled as the mineral nutrient called _____.

55. Nitrogen eventually returns back to the atmosphere because certain bacteria convert the nitrate and ammonia compounds back to _____.

30

56.   Complete the below diagram of the NITROGEN CYCLE by putting in all arrows of interrelationships between organisms and labeling the empty boxes (see Fig 3-16).

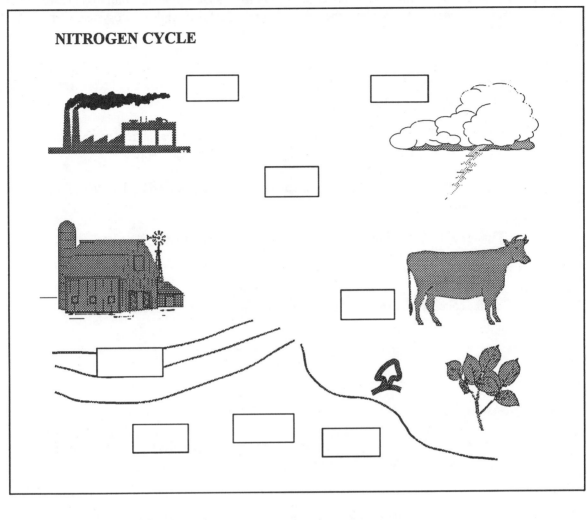

57.   Is it (True or False) that all natural terrestrial ecosystems are dependent on the presence of nitrogen fixing organisms and legumes?

58.   The human agricultural system bypasses nitrogen fixation by using fertilizers containing _____ or _____.

Running on Solar Energy

59.   *For sustainability, ecosystems use sunlight as their source of energy* is the (First, Second, Third) basic principle of ecosystem sustainability.

60.   Identify two benefits derived from solar energy.

      a. _____   b. _____

Prevention of Overgrazing

61.   *For sustainability, the size of consumer populations is maintained so that overgrazing or other overuse does not occur* is the (First, Second, Third) principle of ecosystem sustainability.

62.   Biomass (Increases, Decreases) at each successive trophic level.

63.   The energy available to producers is (Greater, Less) than that available to herbivores, therefore, the biomass of producers will be (Greater, Less) than that of herbivores.

64.   The energy available to herbivores is (Greater, Less) than that available to carnivores, therefore, the biomass of herbivores will be (Greater, Less) than that of carnivores.

65.   Indicate whether the following terms pertain to the first [1], second [2], or third [3] principle of ecosystem sustainability.

    [   ] biomass [   ] recycling [   ] solar energy [   ] nonpolluting [   ] carbon

    [   ] food chains [   ] ammonium and nitrates

66.   Relate the concept of **standing biomass** to that of **sustainable harvest** in Chapter 2.

_____

_____

Implications for Humans

67.   Indicate whether the following human actions are a violation of the first [1], second [2], or third [3] principle of ecosystem sustainability.

    [   ] lack of recycling
    [   ] excessive use of fossil fuels
    [   ] feeding largely on the third trophic level
    [   ] excessive use of fertilizers
    [   ] use of nuclear power
    [   ] using agricultural land to produce meat
    [   ] destruction of tropical and other rainforests
    [   ] nutrient overcharge into marine and freshwater ecosystems
    [   ] production and use of nonbiodegradable compounds

VOCABULARY DRILL

Directions: Match each term with the list of definitions and examples given in the tables below. Term definitions and examples may be used more than once.

Term	Definition	Example (s)	Term	Definition	Example (s)
anaerobic	X		malnutrition	u	
atoms	b		matter	a	
cellulose	v		mineral	f	
chemical energy	n		mixture	e	
compound	d		molecule	c	
elements	b		natural organics	i	
energy	k		organic molecule	g	
entropy	a√		organic phosphate	y	
fermentation	w		overgrazing	z	
First Law of Thermodynamics	0		oxidation	t	
inorganic molecule	h		potential energy	m	
kinetic energy	l		Second Law of Thermodynamics	p, r	
Law of Conservation of Matter	aa		synthetic organics	j	

VOCABULARY DEFINITIONS

a. anything that occupies space and has mass
b. basic building blocks of all matter
c. any two or more atoms bound together
d. any two or more different atoms bound together
e. no chemical bonding between molecules involved
f. any hard crystalline material of a given composition
g. carbon-based molecules which make up the tissue of living organisms
h. molecules with neither carbon-carbon nor carbon-hydrogen bonds
i. compounds produced by living organisms

VOCABULARY DEFINITIONS

j. compounds produced by humans in test tubes
k. the ability to move matter
l. energy in action or motion
m. energy in storage
n. potential energy contained in fuels
o. energy is neither created nor destroyed by may be converted from one form to another
p. in any energy conversion, you will end up with less usable energy that you started with
q. disorder
r. systems will go spontaneously in one direction only: toward increasing entropy
s. process of oxidizing glucose in the presence of oxygen into carbon dioxide and water
t. process of breaking down molecules
u. absence of one or more nutrients in the diet
v. stored form of glucose in plant cell walls
w. partial oxidation of glucose that can occur in the absence of oxygen
x. oxygen-free environment
y. phosphate bound in organic compounds
z. a form of nonsustainable harvest
aa. atoms are not created or destroyed in chemical reactions

VOCABULARY EXAMPLES

a. gases, liquids, solids	j. burning match	s. alcohol
b. NCHOPS	k. a potatoe	t. deep space
c. O2	l. gasoline	u. nucleotides
d. CH4	m. You can't get something for nothing.	v. clear-cut logging
e. muddy water	n. You can't even break even.	w. $C6H12O6 + 6O2 \rightarrow 6CO2 + 6H20$
f. diamond	o. decomposition	
g. NaCl	p. digestion	
h. plastics	q. vitamin deficiency	
i. walking	r. fiber	

1.   Identify 17 elements out of the 108 (Table C-1) that are essential in the structure or function of living organisms? (Your text gave you 16 in Table 3-1.)

The essential elements of life include:_____

_____

If a nonessential element becomes incorporated into a living system (e.g., yourself); what are three potential impacts of this "element insult" on the living system?

2. _____

3. _____

4. _____

Identify from the examples of organization of matter on the right the most simple and the most complex levels.

5. [   ] Most simple

6. [   ] Most complex

a. organisms
b. proteins, carbohydrates, fats
c. atoms
d. cells

7.   What is the carbon source for photosynthesis? _____

8.   What is the energy source for photosynthesis? _____

9.   What is the carbon source for cellular respiration? _____

10.   What is the energy source for cellular respiration? _____

Use the below chemical equations to answer the next series of questions.

a. $6CO_2 + 6H_2O \rightarrow C_6H_{12}O_6 + 6O_2$

b. $C_6H_{12}O_6 + 6O_2 \rightarrow 6CO_2 + 6H_2O$

Which equation:

11.   [   ] is the chemical formula for photosynthesis?

12.   [   ] is the chemical formula for cellular respiration?

13.   [   ] shows glucose as an end product?

14.   [   ] shows the raw materials for photosynthesis as the end product?

15.   [   ] uses a kinetic energy source?

16.   [   ] uses a potential energy source?

17.   How does equation [a] demonstrate the Law of Conservation of Matter?

_____

_____

Associate the below terms with [a] First Law or [b] Second Law of Thermodynamics.

18.   [   ] entropy

19.   [   ] loss

20.   [   ] change

21.   In every energy conversion

a. some energy is converted to heat   b. heat energy is lost
c. lost heat may be recaptured and reused   d. a and b

22.   Consumers must feed on preexisting organic material to obtain

a. nutrients   b. energy   c. both a and b   d. neither a or b - consumers make their own
nutrients and energy

23.   Which of the following is not recycled in natural ecosystems

a. nitrogen   b. carbon   c. energy   d. phosphorus

24.   The carbon atoms in the food you eat will

a. be exhaled as carbon dioxide   b. become part of your body tissues
c. pass through as fecal wastes   d. a, b, and c

25.   In the phosphorus cycle, soil phosphate first enters the ecosystem through

a. plants   b. herbivores   c. carnivores   d. decomposers

26.   Nitrogen fixation refers to

a. releasing nitrogen into the air   b. converting it to a chemical form plants can use
c. repairing broken molecules   d. applying fertilizer

27.   The nitrogen cycle relies on

a. mutualism   b. commensalism   c. parasitism   d. synergism

Indicate whether the carbon [C], nitrogen [N], phosphorus [P], or more than one [M] cycle has been negatively effected by the following human actions.

28.   [   ] abandonment of rotation cropping

29.   [   ] heavy reliance on fossil fuels

30.   [   ] discharges of household detergents into natural waterways

31.   What LAW explains why it is impossible to have a greater biomass at the highest level in a food chain?

        a. Law of Conservation of Matter,   b. First or,   c. Second Law of Thermodynamics

32.   How does this law (your response to item #31) explain why it is impossible to have a greater biomass at the highest level in a food chain.

        _____

        _____

33.   Explain why it makes sense to nurture a recycling mentality in the human system?

        _____

CHAPTER 4

ECOSYSTEMS IN AND OUT OF BALANCE

We have seen that every ecosystem consists of a grouping of plants, animals, and microbes in various roles as producers, consumers, and decomposers being fed upon or feeding upon each other in a complex food web. Overall, we see that the ecosystem is sustained by a cycling of nutrients within the system and a flow of energy through the system. But, what prevents herbivores from eating up all the producers? What prevents carnivores from devouring all their prey? What prevents parasites from killing off the entirety of their host population? What prevents one species from overpopulating and crowding all others? It should be obvious that if any of these things did happen, it would destroy the ecosystem.

Note that we said these events do not *generally* happen! However, when we view humans in the context of the total biosphere, we do see some of these things occurring. Humans are overpopulating and along with their developments, are destroying and crowding out and destroying many species of wildlife and even whole ecosystems. Also humans are such effective predators that we have hunted and are continuing to hunt many species of wildlife into extinction. Further damage is being done by pollution, and by introducing species from one ecosystem into another. These actions, if allowed to continue, can eventually lead to the collapse of the global ecosystem, the biosphere, and, of course, human civilization along with it. Sustaining the biosphere and the human ecosystem within it will depend, first, on our understanding of the checks and balances which underlie the maintenance of natural ecosystems and, second, on our applying such checks and balances to the human system.

STUDY QUESTIONS

1. Predict what would happen to all consumers if the first trophic level was destroyed.

    _____

2. Give two examples of how the concepts of *balance* and *sustainability* are interrelated issues.

    a. _____

    b. _____

Ecosystem Balance is Population Balance

3. Each species found in an ecosystem is represented by a larger, interbreeding, reproducing group called a _____.

4. If an ecosystem is to remain stable over a long period of time, the size of each population must remain relatively _____.

5. In order for a population to remain constant, its reproductive rate must be balanced by an equal _____ rate.

Biotic Potential versus Environmental Resistance

6. Indicate whether the following factors are examples of **biotic potential** [BP] or **environmental resistance** [ER].

   a. [   ] adverse weather conditions   b. [   ] ability to cope with weather conditions

   c. [   ] predators   d. [   ] defense mechanisms

   e. [   ] competitors   f. [   ] reproductive rate

7. If biotic potential is higher than environmental resistance, the population will (Increase, Decrease).

8. Population explosions occur when biotic potential is (Higher, Lower) than environmental resistance.

9. When a population is more or less stable in terms of numbers, biotic potential is (Higher Than, Lower Than, Equal To) environmental resistance.

10. Describe two types of **reproductive strategies** used by plants or animals.

    a. _____

    b. _____

11. Is it (True or False) that the dynamic balances which exist between biotic potential and environmental resistance may be easily upset by humans?

12. Is it (True or False) that when balances are upset, there are population explosions of some species and extinctions of other species?

Density Dependence and Critical Numbers

13. Does environmental resistance (Increase or Decrease) as **population density** increases?

14. How are the concepts of **critical numbers** and **endangered** species related?

    _____

Mechanisms of Population Balance

Predator-prey and Host-parasite Balances

15. Explain how Fig. 4-3 represents an example of *predator-prey balance*.

    _____

16.   If a herbivore population is not controlled by natural enemies, it is likely that the herbivore population will (Increase, Decrease) and the vegetation it feeds on will (Increase, Decrease).

17.   Which of the following groups of organism requires a high degree of adaptation to maintain a balance between environmental resistance and biotic potential? (Check all that apply)

      [ ] predator   [ ] prey   [ ] parasite   [ ] host

18.   Is it (True or False) that a natural enemy can be so effective that it causes the extinction of its prey or host? If you said "True", give an example _____

19.   If individuals are stressed by environmental resistance, indicate whether they would [+] or would not [-] exhibit the following characteristics.

      [   ] Less able to compete successfully with other species.
      [   ] More vulnerable to attack by predators.
      [   ] More vulnerable to attack by parasitic and disease organisms.
      [   ] More likely to attempt to disperse from the area.

When Balances are Absent

20.   Explain how ecosystems got out of balance with the following introductions of exotics.

      a. rabbits in Australia: _____

      b. chestnut blight in America: _____

      c. domestic cats on islands: _____

      d. zebra mussel in Great Lakes: _____

21.   Give an example of an introduction of an exotic that did not effect the balance.

      _____

22.   List the two conditions that enable introduced species to thrive in their new ecosystems.

      a. _____

      b. _____

Territoriality

23.   Is it (True or False) that effective territorial behavior by an animal results in priority use of those resource available in a specific area?

24.   List three options for survival available to individuals unable to claim a territory.

      a. _____   b. _____

      c. _____

Plant-Herbivore Balance

25.   Indicate whether the following activities will maintain [+] or upset [-] the plant-herbivore
      balance.

      [   ] introducing goats on islands
      [   ] killing off herbivore predators
      [   ] confining herbivores in a localized area
      [   ] allowing the herbivores to exceed the carrying capacity of the habitat
      [   ] perpetuating monocultures

Carrying Capacity

26.   Is it (True or False) that there is a finite number of plants or animals that can be supported by an
      ecosystem?

Balances among Competing Plants

27.   Identify three factors that sustain a multiple plant species balance in natural ecosystems.

      a. _____

      b. _____

      c. _____

28.   Indicate whether the following conditions are [+] or not [-] examples of **balanced herbivory**.

      [   ] Plant species unique adaptations to specific microclimates.
      [   ] Introducing plant species that out compete native species.
      [   ] Introducing herbivores that out compete native grazing animals.
      [   ] Introducing carnivores that out compete native carnivores.
      [   ] Introducing carnivores that prey species do not recognize.
      [   ] Development of rubber plantations.

Two Kinds of Population Growth Curves and the Fourth Principle of Ecosystem Sustainability

29.   A population that continues to grow exponentially until it exhausts essential resources and then
      has a precipitous die off demonstrates the _____ curve of population growth.

30.   A population that grows exponentially and then exhibits a leveling off demonstrates the
      _____ curve of population growth.

31.   Animals populations subjected to human disturbances usually demonstrate the _____
      curve of population growth.

32.   The _____ curve of population growth is an indicator of ecosystem nonsustainability.

Ecological Succession

33.   Is it (True or False) that there is an end point in the process of **ecological succession**?

34.   List three examples of ecological succession

a. _____   b. _____   c. _____

Primary Succession

35.   In what order (1 to 4) would the following plant communities appear during primary succession?

[ ] trees   [ ] mosses   [ ] shrubs   [ ] larger plants

36.   How do mosses change the conditions of the area so that it can support larger plants?

_____

Secondary Succession

37.   Suppose that a section of eastern deciduous forest was cleared for agriculture and later abandoned. In what order (1 to 4) would the following species invade the area?

[4] deciduous trees   [1] crabgrass   [3] pine trees   [2] grasses

38.   What prevents hardwood species of the deciduous forest from reinvading the area immediately?

_____

Aquatic Succession

39.   The end result of aquatic succession is a _____ ecosystem.

Succession and Biodiversity

40.   What role does biodiversity play in ecological succession?

_____

Fire and Succession

41.   In which of the below ecosystems can fire be used as a management tool? (Check all that apply)

[ ] grasslands [ ] tropical rainforests [ ] redwood trees   [ ] pine forests

42.   Indicate whether the following events are [+] or are not [-] the results of controlled ground fires.

[ ] removal of dead litter
[ ] permanent harm of mature trees

[   ] seed release of some tree species
[   ] destruction of wood-boring insects
[   ] crown fires
[   ] loss of human lives and property

The Fourth Principle of Ecosystem Sustainability

43.   Write out the fourth principle of ecosystem sustainability.

_____

Implications for Humans

44.   The human population is exploding because we have effectively

         a. increased our biotic potential. (True or False)
         b. decreased our environmental resistance. (True or False)

45.   The human population growth pattern presently resembles a _____ curve.

46.   List three ways in which humans are causing the upset of natural ecosystems.

         a. _____

         b. _____

         c. _____

Term	Definition	Example (s)	Term	Definition	Example (s)
aquatic succession			host specific		
balanced herbivory			monoculture		
biotic potential			natural enemies		
carrying capacity			population explosion		
climax ecosystem			population density		
critical number			primary succession		
ecological restoration			recruitment		
ecological succession			replacement level		
environmental resistance			reproductive strategies		
exponential increase			secondary succession		
fire climax			territoriality		

VOCABULARY DEFINITIONS

a. number of offspring a species may reproduce under ideal conditions

b. number of offspring that survive and reproduce

c. the way species achieve recruitment in their populations.

d. a doubling of each generation

e. the result of exponential increase

f. biotic and abiotic factors that tend to decrease population numbers

g. number of offspring required to replace breeding population

h. number of individuals per unit area

i. the minimum population base (number) of a species

j. predators and parasites that control population size

k. defending territory against encroachment of others of the same species

VOCABULARY DEFINITIONS
l. maximum population a given habitat will support without the habitat being degraded over time
m. growth of a single species over a wide area
n. parasites that attack only one species and close relatives
o. balance among competing plant populations maintained by herbivores
p. the orderly transition from one biotic community to another
q. a stage of equilibrium between all species and the physical environment
r. succession in an area that has not been previously occupied
s. succession in an area that has been previously occupied
t. replacement of aquatic by terrestrial communities
u. ecosystems that depend on the recurrence of fire to maintain existing balance
v. reclaiming natural ecosystems

VOCABULARY EXAMPLES
a. live births, eggs laid, seeds in plants
b. seeds that sprout and produce more seeds
c. produce massive numbers of young
d. $2 \rightarrow 4 \rightarrow 8 \rightarrow 16 \rightarrow 32$
e. lack of food, space, and mates
f. number of offspring required to replace breeding population
g. 2 deer/acre
h. endangered species
i. predators and parasites
j. scent marking
k. upper limit on S-shaped population growth curve
l. a field of cotton
m. host specific: dog tape worm
n. the opposite of monoculture
o. bare rock $\rightarrow$ moss $\rightarrow$ herbaceous plants $\rightarrow$ shrubs $\rightarrow$ trees
p. deciduous forest

VOCABULARY EXAMPLES
q. deciduous forest → monoculture → grassland → shrubs → pine forest → deciduous forest
r. lake → marsh → grassland
s. grasslands and pine forests
t. wetlands

SELF-TEST

Indicate whether the following factors are examples of biotic potential [BP] or environmental resistance [ER].

1. [   ] lack of suitable habitat   2. [   ] disease   3. [   ] reproductive rate

4. [   ] ability to migrate   5. [   ] + adaptability   6. [   ] + competition

Identify whether the following traits are characteristic of [J] or [S] population growth curves.

7. [   ] biotic potential = environmental resistance

8. [   ] biotic potential > environmental resistance

9. [   ] population increases and dies off in rapid sequence

10. [   ] violates the Fourth Principle of Ecosystem Sustainability

11. [   ] the human population growth curve is an example

12.   What concept concerning population numbers is the direct opposite of carrying capacity

_____

13.   In what order (1-4) would the following plant communities appear during primary succession?

    [   ] trees   [   ] mosses   [   ] grasses   [   ] shrubs

14.   In what order (1-4) would the following plant communities appear during secondary succession?

    [   ] grasses   [   ] pine trees   [   ] herbaceous plants   [   ] deciduous trees

15.   What ecological factors define primary versus secondary succession?

_____

16.   In which of the following ecosystems can fire not be used as a management tool?

    a. grasslands   b. coniferous forests   c. hardwood forests   d. a and b

17. Which of the following is not an intended function of ground fires?

   a. removal of dead litter   b. seed release of some tree species
   c. starting crown fires   d. destruction of wood-boring insects

18. Aggressive defense of habitat is called

   a. biotic potential   b. succession   c. territoriality   d. the niche

19. The process whereby the growth of one community causes changes that make the environment more favorable to a second community and less favorable to the first is called

   a. biotic potential   b. succession   c. territoriality   d. the niche

20. Which of the following is not a characteristic of climax ecosystems?

   a. have maximum species diversity
   b. have different dominant plant forms
   c. are self-perpetuating
   d. will maintain themselves regardless of climatic, abiotic, or biotic changes

   Which of the following human actions will [-] and will not [+] cause upsets of natural ecosystems? (Note: the following list expands on that given in the text)

21.   [   ] increasing population size

22.   [   ] decreasing the gross national product

23.   [   ] supporting conservation organizations

24.   [   ] introducing species from one ecosystem to another

25.   [   ] eliminating natural predators

CHAPTER 5

ECOSYSTEMS: ADAPTING TO CHANGE-OR NOT

At this point in your study of how ecosystems work, you should be becoming aware of the fact that even though organisms attempt to maintain a dynamic balance between biotic potential and environmental resistance - the struggle is everlasting. This is the case because ecosystems are in a constant state of change. These changes are the result of subtle or dramatic alterations in the abiotic or biotic characteristics of the ecosystem. However, whether the change is subtle (e.g., slight change in pH) or dramatic (e.g., turning a grassland into a corn field), each species has only three response alternatives to these changes, i.e., adapt, move, or die. The choice of which alternative is used will be based on the information available, or not, in the gene pool of a species.

Chapter 5 provides an overview of how genetics plays the major role in adaptation, change, or extinction. As soon as humans realized how genetics plays this role, we developed technologies that allowed us to manipulate (e.g., hybridization and genetic engineering) the gene pools of selected species. We now realize that these manipulations of gene pools also required the development of safeguards against natural selection. For example, hybrid seed corn was developed to produce lots of corn, not to combat environmental resistance. In order to keep hybrid seed corn alive and to combat environmental resistance, farmers need to provide tremendous amounts of water, energy, and nutrients. Further examples of how the gene pool of a species hinders human efforts to control environmental resistance will be discussed in later chapters on pest control.

STUDY QUESTIONS

Selection by the Environment

1. Identify two ways in which changes in the **gene pool** might occur.

    a. _____   b. _____

2. Is it (True or False) that all individuals in a population reproduce at the same rate?

3. Is it (True or False) that survival automatically leads to reproduction?

Change Through Selective Breeding

4. Selective breeding requires a (Repetitive, One-time) process in order to develop the desired trait.

5. Selective breeding (a. Increases, b. Decreases, c. both a and b) certain alleles in the gene pool of a population.

6. All the different breeds of dogs are derived from the (Same, Different) wild dog gene pool.

7.   It is genetically logical to mate a Great Dane with a Chihuahua. (True or False)

Change Through Natural Selection

8.   In nature, every generation of every species is subjected to an intense selection of
        _____ and _____.

9.   New alleles that provide for or enhance a trait that aids in survival and reproduction will
        (Increase, Decrease) in the gene pool of the population.

Adaptations to the Environment

10.   Indicate whether the following adaptations would [+] or would not [-] support the survival and
        reproduction of organisms.

        a. [   ] Coping with climatic and other abiotic factors.
        b. [   ] Obtaining food and water.
        c. [   ] Escaping from predation.
        d. [   ] Ability to attract mates.
        e. [   ] Ability to migrate to adjacent areas.

11.   Give an example of (a to e, question 10, see Fig. 5-4).

        a. _____

        b. _____

        c. _____

        d. _____

        e. _____

Selection of Traits and Genes

12.   Provide an example of a trait related to:

        a. physical appearance _____

        b. tolerance _____

        c. behavior _____

        d. metabolism _____

13.   The hereditary component of all traits of all organisms is encoded in _____
        molecules.

14. The collection of all the DNA molecules or genes within a cell of an organism is called its _____ makeup.

15. Relate DNA to structural, enzymatic, and hormonal proteins to physical, metabolic, and growth traits in the below diagram (see Fig. 5-5 for an example of this exercise)

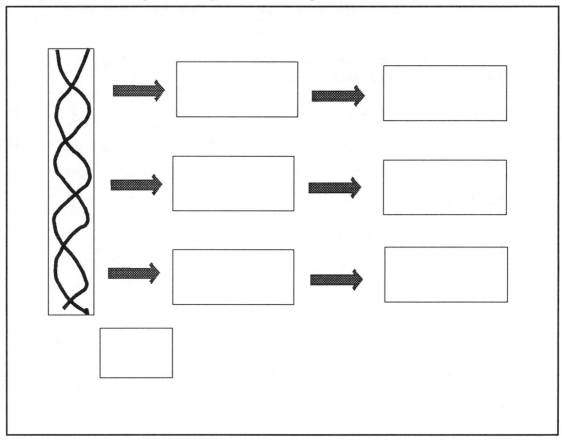

Genetic Variation and Gene Pools

16. The genetic makeup of nearly all organisms consists of (One, Two, Three, Four) complete set(s) of genes or (One, Two, Three, Four) pair(s).

17. Indicate whether each of the below cell types would have one [1] or two [2] alleles of a gene pair.

    [   ] body cells   [   ] sperm cell   [   ] egg (ovum)   [   ] fertilized egg

18. A sperm cell entering an egg is called the process of _____.

19. Each fertilization will result in a different combination of _____ and this combination will be different from that of either parent.

20. Explain why it is more likely for males to contract sex chromosome related diseases than females. _____

21. What is the genetic difference between identical twins and **clones**?

_____

Mutations-the Source of New Alleles

22. Explain what a **mutation** is in terms of the DNA molecule.

_____

23. Is it (True or False) that mutations are rarely beneficial?

Changes in Species and Ecosystems

24. Natural selection on gene pools is a (Quick, Slow) process.

Speciation

25. Rank (1 to 4) the below events in terms of their order of occurrence in the process of speciation.

   [   ] two or more species developing from one
   [   ] original species gene pool separated
   [   ] long-term loss of interbreeding potential
   [   ] single species gene pool

26. If a species moves to a new location where biotic and/or abiotic conditions are somewhat different, which of the following events could happen?

   a The migrant population may die out. (True or False)
   b. Some individuals may survive and reproduce. (True or False)
   c. There will be selection pressure for alleles that enhance survival. (True or False)
   d. Alleles in the gene pool of the migrant population will become different from those in the gene pool of the original population. (True or False)
   e. Speciation could occur. (True or False)

27. If natural selective pressures are for animals with long hair and short legs, then the majority of the population will have (long, short) hair and (long, short) legs.

28. Give one example of migration and speciation that occurred on the Galapagos Islands.

   _____

Developing Ecosystems

29. (All, A Few) species within an ecosystem undergo simultaneous adaptation and change through natural selection.

30. Every species exists and can only exist in the context of:

   a. producers   b. consumers   c. decomposers   d. the whole ecosystem

31.   Identify the original organisms that occupied the grazing niche:

        a. on Galapagos Islands: _____

        b. in Brazil: _____

        c. in North America: _____

        d. in Australia: _____

32.   Explain why the natural selection process chose different species to occupy the grazing niche in
        each of the areas listed in question #31.

        _____

33.   Is it (True or False) that if you introduced rabbits into Australia, they would become the prime
        herbivores?

34.   Is it (True or False) that if you introduced cattle into Australia, they would become the prime
        herbivores?

How Rapid is Evolution?

35.   List the stepwise process of evolution.

        a. Step 1: _____

        b. Step 2: _____

        c. Step 3: _____

        d. Step 4: _____

36.   Selective pressures acting on the gene pools of all species usually lead to development of a
        _____ ecosystem.

Limits of Change

37.   List the three options available to each species that is ill-adapted to a new ecosystem.

        a. _____ b. _____ c. _____

38.   Adaptations are:

        a. present in a species gene pool prior to environmental selection.
        b. the result of mutations occurring after the environment selects for a particular trait.
        c. variations on pre-existing genetic themes in the species' gene pool.
        d. a and c are both correct

39. What two criteria must be met in order for survivors to adapt to a new environmental condition?

   a. _____   b. _____

40. Indicate whether the following are [+] or are not [-] criteria for populations of individuals to survive new conditions.

   [   ] High degree of genetic variation in the gene pool
   [   ] Wide geographic distribution
   [   ] High reproductive capacity
   [   ] Being small

41. Compare cockroaches and elephants in terms of whether they do [+] or do not [-] meet the above (question #40) criteria for survival.

   Cockroaches

   [   ] High degree of genetic variation in the gene pool
   [   ] Wide geographic distribution
   [   ] High reproductive capacity
   [   ] Being small

   Elephants

   [   ] High degree of genetic variation in the gene pool
   [   ] Wide geographic distribution
   [   ] High reproductive capacity
   [   ] Being small

42. (Cockroaches or Elephants) have the best chance of adapting to new environments?

43. (Cockroaches or Elephants) are more likely to adapt to changes caused by humans?

Global Changes

44. The earth's crust is (thick, thin) and in a constant state of (movement, inertia).

45. The floating slabs of rock that make up the earth's crust are called _____ plates.

Tectonic Plates

46. Provide four sources of evidence that the earth's crust is in a constant state of movement.

   a. _____   b. _____

   c. _____   d. _____

47. Define "Pangea":

   _____

48.   List three ways that movement of the earth's crust effects climate.

      a. _____

      b. _____

      c. _____

An Overview of Evolution

49.   If the entire geological history of Earth was made equivalent to one year, when did humans:

      a. appear? _____   b. develop agriculture? _____

      c. develop technological knowledge? _____

The Human Animal

50.   Over 98% of human DNA is identical to _____ DNA.

51.   List three characteristics of the human animal that make them adaptively different from its closest genetic relative.

      a. _____

      b. _____

      c. _____

52.   Relate the Fourth Principle of Ecosystem Sustainability to:

      a. endangered species: _____

      b. agriculture: _____

      c. biotechnology: _____

      d. medicine: _____

Directions: Match each term with the list of definitions and examples given in the tables below. Term definitions and examples might be used more than once.

Term	Definition	Example (s)	Term	Definition	Example (s)
adaptation			mutation		
alleles			natural selection		
biological evolution			selective breeding		
differential reproduction			selective pressures		
DNA			speciation		
gene			tectonic plates		
gene pool			trait		
genetic variation					

VOCABULARY DEFINITIONS

a. total genetic differences that exist among individuals within a population

b. total number of genes within a population

c. individuals reproducing more than others in a population

d. change in the gene pools of species over course of generations

e. a human-induced form of differential reproduction

f. environmental resistance factors that determine survival and reproductive rates

g. selection and modification of species' gene pools through natural forces

h. any particular characteristic of an individual

i. a process that results in two or more species from one

j. huge slabs of rock floating on the molten Earth's core

k. the ability to form balanced relationships with biotic community and abiotic environment

l. the blueprint of the total genetic make up of organisms

m. a segment of DNA that codes for one particular protein

n. different forms of a gene

o. inheritable change in the DNA molecule

VOCABULARY EXAMPLES	
a. races of humans	g. Galapagos finches
b. 50 to 100 K for humans	h.: San Andreas fault
c. great danes and dachshunds	i. a giraffes long neck
d. fish → amphibians → reptiles → mammals	j. double alpha-helix
e. every factor of environmental resistance	k . nucleotide sequence
f. blue eyes, dark skin	l. altered nucleotide sequence

SELF-TEST

Indicate whether the following terms or descriptions refer to traits [T] or genes [G].

1. [   ] DNA segment   2. [   ] physical appearance   3. [   ] alleles

Check any of the following events that increase genetic diversity in a species gene pool.

4. [   ] sexual reproduction   5. [   ] selective breeding   6. [   ] recombination

7.   Which of the following statements is not true concerning DNA molecules? DNA molecules are

   a. the site of genetic information within organisms.
   b. found only in sperm or egg cells.
   c. capable of being influenced by genetic engineering.
   d. one of the locations for mutations to occur.

8.   How many alleles of a gene pair would normally be found in a sperm cell?

   a. one   b. two   c. three d. four

9.   How many alleles of a gene pair would normally be found in a skin cell?

   a. one   b. two   c. three d. four

Complete each of the following statements with the terms MORE or LESS.

10.   More selective pressure leads to _____ reproduction.

11.   Less adaptation to abiotic and biotic factors leads to _____ reproduction.

12.   Explain why more survival does not automatically lead to more reproduction.

   _____

   _____

13.   Which of the following statements is not correct regarding selective breeding?

        a. It is a one-time event.
        b. It increases genetic diversity in the gene pool of a population.
        c. It decreases genetic diversity in the gene pool of a population.
        d. It is a human-induced mechanism to control natural selection.

14.   Mutations that are passed on from generation to generation are called:

        a. beneficial   b. neutral   c. lethal   d. a and b

15.   Explain why the giant tortoises were so successful in occupying the herbivore niche on the Galapagos Islands:

        _____

        _____

        Which of the following adaptations gave goats the **greatest** advantage over giant tortoises as the prime herbivores on the Galapagos Islands. (Check all that apply and defend selections).

        <u>Adaptations</u>   <u>Defense of Answer</u>

16. [   ] coping with abiotic factors   _____

17. [   ] obtaining food   _____

18. [   ] escaping from predation   _____

19. [   ] ability to migrate   _____

        Identify the criteria for species survival in the list given below and then indicate whether this is a criterion of the human population. (Check all that apply)

                                                         Criteria Present in
        <u>Criteria for Survival</u>   <u>Human Population</u>

20. [   ] Genetic diversity in gene pool   23. [   ]

21. [   ] Wide geographic distribution   24. [   ]

22. [   ] Rapid reproductive rate   25. [   ]

Identify those conditions that represent the human system (check all that apply) and then predict whether the listed conditions would promote [+] or prevent [-] the future survivability of the human species.

Conditions   Survivability

26. [   ] causing rapid abiotic/biotic changes   30. [   ]

27. [   ] monoculture development   31. [   ]

28. [   ] sustainable agriculture   32. [   ]

29. [   ] preserving biodiversity   33. [   ]

34.   Explain the contradiction you should have noticed in your responses to items 20 to 25 and items 26 to 33.

       _____

       _____

35.   Identify a major force behind climate changes: _____

CHAPTER 6

THE GLOBAL HUMAN POPULATION EXPLOSION:
CAUSES AND CONSEQUENCES

Over the past several decades, the human population has been growing at a rate of 70-90 million people per year. This means that an additional number of people equivalent to about a third of the United State's population will be competing for food, clean water, safe shelters, health care services, jobs, and an education. Unfortunately, many of these new arrivals are born in situations of absolute poverty. But this is only half the tragedy! The other half is that the Earth's resources are being depleted, undercut, and destroyed in order to sustain high fertility rates in some countries and affluent consumption in others. These trends can only lead to upset and collapse of the entire biosphere with unprecedented suffering for all.

At what point may the collapse occur? What is the carrying capacity of the Earth for the human species? We will be able to answer the latter question when the collapse occurs. Is the human species increasing environmental resistance or biotic potential for itself and all other species?

Study shows that it is not incidental that high birth rates and, hence, population growth are connected with poverty. Poverty and high birth rates are entwined in a vicious cycle. For example, the average American woman who lives long enough winds up with a total of 14 children, grandchildren, and great-grandchildren. Average West German women, only 5. Average African woman, **258**. To break this cycle, we must understand it and attack it from the side of poverty as well as birth rate itself. Conveying this understanding and presenting the avenues of attack are objectives of this chapter.

STUDY QUESTIONS

The Population Explosion and Its Cause

The Explosion

1. Chart the progress of human population growth from 1830 to that projected for 2046.

Year	Billions of People	Years
a. 1830		
b. 1930		
c. 1960		
d. 1975		
e. 1987		
f. 1998		
g. 2009		
h. 2020		
i. 2033		
j. 2046		

Reasons for the Explosion

2.   Why was the human population in a dynamic balance prior to the 1800s?

        _____

3.   List four factors that increased the recruitment rate of the human population after the 1800s.

        a. _____   b. _____

        c. _____   d. _____

4.   Human population growth is expected to level off at _____ billion.

Different Worlds

Rich Nations and Poor Nations

5.   The countries of the world fall into three major economic categories (A. high-income, highly
        developed; B. middle-income, moderately developed; and C. low-income). Identify these
        categories with examples of countries or regions that are in each category (see Fig. 6-3).

        [   ] United States, Japan, Western Europe
        [   ] Latin America, East Asia
        [   ] East and Central Africa, India, China

6.   Developed countries have about _____ percent of the world's population, but they control
        about _____ percent of the world's wealth.

7.   Developing countries have _____ percent of the world's population, but control only
        _____ percent of the world's wealth.

8.   List six conditions of "absolute poverty".

        a. _____   b. _____   c. _____

        d. _____   e. _____   f. _____

Population Growth in Rich and Poor Nations

9.   Identify the countries that have the shortest and longest doubling times (see Tables 6-2):

        a. _____   b. _____

10.   Over the past few decades, the fertility rates in developed countries have (Increased, Decreased)
        and (Increased, Decreased) in developing countries.

Different Populations Present Different Problems

11. Match the characteristics below with [a] developed or [b] developing countries.

   [b] high total fertility rates
   [a] low total fertility rates
   [b] short doubling times
   [a] high consumptive lifestyles
   [a] long doubling times
   [b] annual rates of increase ≥ 2%
   [a] annual rates of increase < 1%
   [b] population growth is most intense
   [b] poverty is most intense
   [b] environmental degradation is most intense
   [a] eating high on the food chain
   [a] demand for tropical hardwoods
   [a] demand for exotic wildlife

12. The negative impacts of high consumptive lifestyles may be moderated by a factor called _____environmental_ **regard**.

13. What three conditions must be met in order to achieve a sustainable population?

   a. _stabilize population_   b. _decrease consumptive lifestyle_

   c. _increase environmental regard_

Environmental and Social Impacts of Growing Populations and Influence

The Growing Populations of Developing Nations

14. List five ways people in developing nations are attempting to resolve population growth and land availability.

   a. _____   b. _____

   c. _____   d. _____

   e. _____

Subdividing Farms and Intensifying Cultivation

15. Indicate whether the following are [+] or are not [-] environmental conditions resulting from attempts to increase agriculture in developing countries.

   [   ] decrease in per capita agricultural land
   [   ] attempt agriculture on marginal land
   [   ] soil erosion
   [   ] loss of topsoil

[ ] water pollution
[ ] loss of soil productivity
[ ] deforestation

Opening Up New Lands for Agriculture

16.   What is the fundamental problem with attempting to open new lands for agriculture in developing countries? _____

Migration to Cities

17.   Indicate whether the following are [+] or are not [-] environmental/social conditions resulting from people in developing countries attempting to seek a better life by moving from the country to cities.

[ ] high population densities
[ ] lack of utilities and social services
[ ] polluted water
[ ] disease
[ ] high crime rates

Illicit Activities

18.   Identify two sources of illicit income that people in developing countries might turn to in order to earn enough money to survive.

a. _____   b. _____

19.   Do the majority of their customers come from (Developed, Developing) countries?

Emigration/immigration

20.   Many people in developing countries see (Emigration, Immigration) as their best future for a brighter future.

21.   Developed countries are placing (More, Fewer) restrictions on immigration.

Impoverished Women and Children

22.   Is it (True or False) that women in developing countries have access to an adequate social welfare system?

23.   Is it (True or False) that cities in developing countries have high numbers of stray or abandoned children?

Effects of Increasing Affluence

24.   Is (True or False) that the affects of affluence are all negative?

25. The United States contains _____ percent of the world's population and is responsible for _____ percent of the global emissions of carbon dioxide.

26. Explain why the cities in developing countries are more polluted (air and water) than cities in developed countries?

_____

Summing Up

27. TAKE A STAND! What is your position [a. agree, b. neutral, c. disagree] concerning the following population issues?

[ ] Further population growth will be beneficial.
[ ] Our technology will solve any problems associated with population growth .
[ ] Even at a population of 12 billion, we can all enjoy an affluenct lifestyle.

Dynamics of Population Growth
Population Profiles

28. Population profiles can be used to project future:

a. _____   b. _____

29. Indicate whether the following fertility rates will lead to an increase [+] decrease [-], or stability [0] in population growth.

[−] Fertility rate < 2
[0] Fertility rate = 2
[+] Fertility rate > 2

30. Which of the above fertility rates is known as **replacement fertility**? =2 _____

31. Match the below shapes of population profiles with the fertility rates given in question #29.

a. Column-shaped: Fertility rate = __2__
b. Pyramid-shaped: Fertility rate = __>2__
c. Upside down pyramid: Fertility rate = __<2__

32. The population pyramid of Denmark would be _____ shaped.

33. The population pyramid of Kenya would be _____ shaped.

Population Projections

34. Identify two population projections provided in population profiles.

a. _____   b. _____

Based on the population profiles for Denmark and Kenya, predict whether **MORE** or **LESS** of the following goods and services will be needed over the next 20 years.

Service	Denmark	Kenya
Schools	35. _____	39. _____
Facilities for old people	36. _____	40. _____
New housing units	37. _____	41. _____
Utility services	38. _____	42. _____

Population Momentum

43. Given the population momentum of Kenya their populations will continue to grow for _____ to _____ years after achieving replacement fertility levels.

Changing Fertility Rates and the Demographic Transition

44. What has been the observed trend in economic development and fertility rates?
_____

45. Subtracting the crude death rate from the crude birth rate (CBR-CDR) and then dividing by 10 gives the annual rate of _____.

46. Doubling time for a population may be found by dividing the annual rate of increase into (7, 70, 700).

47. Calculate the annual rate of increase and doubling time for the following countries.				
Country	Crude Birth Rate	Crude Death Rate	Rate of Increase	Doubling Time
Kenya	46	7		
Mexico	29	6		
United States	17	9		
Denmark	12	12		

48. The goal of each country should be to have a (Long, Short) doubling time.

49. The demographic transition refers to changes in infant _____ and _____ rates which tend to parallel development.

50. There are (One, Two, Three, Four) major phases in the demographic transition?

51. In the table below, indicate whether the variables given after each phase are high [+], declining [-], or stable [0].

Phase	Birth Rate	Death Rate	Population
I	[ ]	[ ]	[ ]
II	[ ]	[ ]	[ ]
III	[ ]	[ ]	[ ]
IV	[ ]	[ ]	[ ]

52. Developed nations are in phase _____ .

53. Developing nations are in phase _____ .

VOCABULARY DRILL

Directions: Match each term with the list of definitions and examples given in the tables below. Term definitions and examples might be used more than once.

Term	Definition	Example (s)	Term	Definition	Example (s)
absolute poverty	l	a	doubling time	i	e
age structure	g	h	environmental regard	j	f
crude birth rate	e	b	natural increase	h	g
crude death rate	f	b	population profile	o	h
demographer	b	h	population momentum	c	i
demographic transition	d	c	replacement fertility	a	j
developed countries	m	d	total fertility rate	x	x
developing countries	n	l			

VOCABULARY DEFINITIONS

a. fertility rate that will just replace the population of parents

b. collecting, compiling, and presenting information regarding populations

c. continued population growth after fertility rate = replacement rate

d. the gradual shift from primitive to modern condition that is correlated with development

e. number of births per 1000 of the population per year

65

VOCABULARY DEFINITIONS
f. number of deaths per 1000 of the population per year
g. relative proportion of people in each age group
h. net increase or loss of population per year
i. the number of years it will take a population growing at a constant percent per year to double
j. factors that moderates environmental impacts of affluent lifestyles
k. average number of children each woman has over her lifetime
l. a condition of life so limited by malnutrition, iliteracy, disease, high infant mortality, and low life expectancy as to be beneath any defintion of human decency.
m. high income nations
n. middle to low income nations
o. the number of people at each age for a given population

VOCABULARY EXAMPLES	
a. 90% of the people in developing countries	g. (CBR - CDR) ÷ 10
b. Table 6-2	h. Fig. 6-11
c. Fig. 6-18	i. 50 to 60 years in Kenya
d. Kenya	j. 2.0
e. 18 to 1155 years	k. 1.3 to 6.7
f. recycling + conservation + pollution control	l. United States

SELF-TEST

1.   Draw a graph showing the change in human population size on the Y axis from historical times through the present and projected into the future on the X axis.

Y axis

X axis

Identify the following countries or regions as [a] developed or [b] developing countries.

2. [   ] Latin America   3. [   ] Japan   4. [   ] Central Africa   5. [   ] United States

6.   Developed countries have about _____ percent of the world's population but control about _____ percent of the world's wealth.

7.   Calculate the people to resource ratio for developed countries in units of resource per person. _____

8.   Developing countries have about _____ percent of the world's population but control only _____ percent of the world's wealth.

9.   Calculate the people to resource ratio for developing countries in units of resource per person.

        _____

        Assume that people living in HDC's reduced their resource consumption by 50% and donated the additional resource to LDC's.

10.   Calculate the new people to resource ratio for developed countries in units of resource per person. _____

11.   Calculate the new people to resource ratio for developing countries in units of resource per person. _____

        Explain the results of your calculations in items 10 and 11 in terms of:

12.   Impact on developed countries: _____

13.   Impact on developing countries: _____

Calculate the annual rate of increase and doubling time for the countries listed below.				
Country	CBR	CDR	Rate of Increase	Doubling Time
14. Developed	14	9	2.4	30
15. Developing	34	10	0.5	137

Indicate whether the following consequences of exploding population are the results of high fertility rates in [a] developing countries, the affluence of [b] developed countries, or [c] both

16. [   ] population pressure in the countryside   17. [   ] pressures on forests and endangered species

18. [   ] global pollution   19. [   ] high emigration rates

20.   Give one example of how population profiles can be used.

   _____

21.   In the phases of demographic transition, having stable birth and death rates is generally seen in

       Phase (I, II, III, IV).

22.   A population profile shows

       a. standard of living.   b. causes of death.

       c. numbers of people in each 5-year age group. d. factors that control fertility rates.

23.   The most significant factor(s) that determine how fast a population grows is (are)

       a. age structure of the population.   b. total fertility rate.

       c. infant and childhood mortality rates.   d. all of the above

24.   The population profile of a developed country has the shape of a

       a. column.   b. pyramid.   c. upside down pyramid.   d. football on its point.

25.   Why would Kenya have to wait 50 years to experience a population decrease?

   _____

CHAPTER 7

ADDRESSING THE POPULATION PROBLEM

Chapter 6 gave you a basic understanding of those factors responsible for explosive human population growth in some countries and a stable, or nearly stabilized, growth in other countries. You may recall that the fundamental resolution to population growth is to control fertility rate. However, in many countries, particularly the less developed nations, there are strong socioeconomic factors that preclude a lowered fertility rate. People in less developed countries may want to have fewer children, but they cannot afford to given the need for a larger family work force, the dependence on children to support their parents when they grow old, and social pressures to have large families. Sometimes, just plain ignorance of family planning leads to high fertility rates.

The brutal realities of the impact of explosive population growth on people's quality of life are the focus of Chapter 7. It is important for us to recognize that uncontrolled human population growth and the associated environmental degradation affects, either directly or indirectly, all people and nations of the world. There is no difference between a person living in India or the United States regarding their basic needs, i.e., food, water, housing, health, an education, and a job. People living in highly developed countries take these basic needs for granted because they are so readily available. People in less developed countries will attempt to achieve these qualities of life regardless of the environmental consequences, (e.g., cutting down the rain forests to farm or graze cattle). Chapter 7 addresses the population problem in terms of improving the lives of people, reducing fertility, and protecting the environment.

STUDY QUESTIONS

Population and Development

1. Interpret Fig. 1 in terms of what Thomas Malthus pointed out in 1798.

   _____

   _____

   _____

   _____

2. What two considerations were not included in the Malthusian interpretation of Fig. 1?

   a. _____

   b. _____

FOOD PRODUCTION

POPULATION GROWTH

TIME

Figure 1

3. What are two basic schools of thought on the issue of population and development?

a. _____ b. _____

Promoting Development of Low-Income Nations

4. What are the expected sequence of events from promoting development in low-income nations?

BETTER JOBS → BETTER _____ → HIGHER STANDARD OF _____ → EXPANDED _____ → FOSTERING _____, _____, AND _____ → LOWER _____ RATES.

5. The world bank has influenced the course of development in poor nations for the past _____ years.

Successes and Failures of the World Bank

6. Score the world bank's successes [+] and failures [-] in promoting development in low-income nations using the following list.

[ ] increased GNP for some countries
[ ] social progress in some countries
[ ] declining fertility rates in some countries
[ ] achievement of replacement level fertility rates in some countries
[ ] the number of people living in absolute poverty
[ ] electric generating facility in India
[ ] large cattle operations in Latin America
[ ] mechanized plantations for growing cash crops

Development or Exploitation of the Poor?

7. Indicate whether the following are [+] or are not [-] the working conditions in industrial plants in developing nations.

[ ] union-protected workforce
[ ] low wages and no fringe benefits
[ ] enforced pollution regulations
[ ] enforced child labor laws

8. Indicate whether the following are [+] or are not [-] benefits of World Bank loans applied to agriculture in developing countries.

[ ] higher production per acre
[ ] crop diversity
[ ] year-round food availability
[ ] direct farmer-to-consumer exchange
[ ] maintains the traditions of subsistence farming

The Debt Crisis

9.   Is the national debt obligation of developing nations (Growing, Shrinking)?

10.   What happens when a developing nation cannot even pay the interest on their loan?

       _____

11.   What are three strategies used by developing nations to make even partial payment on their debt?

       a. _____   b. _____

       c. _____

12.   How is the debt crisis in developing nations "the essence of nonsustainability"?

       _____

13.   Give three examples of the large-scale centralized projects funded by the World Bank.

       a. _____   b. _____

       c. _____

14.   Is it (True or False) that large-scale centralized projects have alleviated absolute poverty conditions?

Reassessing the Demographic Transition

15.   Is it (True or False) that fertility rates in industrialized countries have declined with development?

Factors Influencing Family Size

16.   Indicate whether the following social and economic factors would **most** determine the family size of [A] developing, [B]developed, [C] both, or [D] neither countries.

       [   ] children: an economic asset or liability
       [   ] old-age security
       [   ] importance of education
       [   ] status of women
       [   ] religious beliefs
       [   ] infant and childhood mortality rates
       [   ] availability of contraceptives

Conclusions

17.   Which of the following social events prevented a population explosion in industrialized societies? (Check all that apply)

[   ] reduction in infant and childhood mortality
[   ] availability of jobs in the cities
[   ] lower fertility rates
[   ] strict control of immigration

A New Direction for Development

18.   Identify those factors that are needed to break the cycle of poverty, high fertility, and environmental degradation. (Check all that apply)

[   ] good health and nutrition for mothers and children
[   ] old-age security apart from children
[   ] educational opportunities for women
[   ] availability of contraceptives
[   ] career opportunities
[   ] rights of women
[   ] improving resource management

19.   Indicate whether the following practices are [+] or are not [-] characteristic of family planning services and information.

[   ] Counseling on reproduction, hazards of STDs, and contraceptive techniques.
[   ] Supplying contraceptives.
[   ] Counseling on pre- and post-natal health of mother and child.
[   ] Counseling on the health advantages of spacing children.
[   ] Promoting as many abortions as possible.
[   ] Nursing a baby for as long as possible.
[   ] How to avoid unwanted or high-risk pregnancies.

20.   Which of the following elements are confounding the process of helping women in developing countries to lower their fertility? (Check all that apply)

[   ] inadequate government funding
[   ] lack of access to and cost of contraceptives
[   ] inadequate sources of information on contraceptives

21.   What three factors preclude most people in developing countries from obtaining a loan?

a. _____   b. _____

c. _____

22.   What is the Gumman Bank doing to provide start-up money for people in developing countries?

_____

23.   Who are the preferred customers of the Grumman Bank? _____

24.   Projects funded by the Grumman Bank:

   a. do not upset the existing social structure. (True or False)
   b. involve a high degree of training or skill. (True or False)
   c. utilize local resources. (True or False)
   d. utilize centralized workplaces. (True or False)
   e. allow individuals to develop self-reliance. (True or False)

25.   Indicate whether the following are examples of projects funded by the World [W] or Grumman [G] bank.

   [   ] Textile mill
   [   ] Handloom
   [   ] Large apartment complex
   [   ] Handmade bricks
   [   ] More efficient wood stoves

Putting it All Together

26.   What are the expected sequence of events from breaking the cycle of poverty, high fertility, and environmental degradation in low-income nations?

   BETTER HEALTH → ECONOMIC ___productivity___ → BETTER ___education___ → DELAYED ___marriage___ → FEWER ___children___

The Cairo Conference

27.   State the conceptual framework of all goals established in the 1994 ICPD Program of Action.

   _____

   _____

28.   How will the Program of Action be funded?

   _____

NO VOCABULARY DRILL FOR THIS CHAPTER!

SELF-TEST

Indicate whether the following social events have happened in [a] developed, [b] developing, of [c] both countries.

1.   [   ] reduction in infant and childhood mortality
2.   [   ] availability of jobs in the cities
3.   [   ] lower fertility rates

Indicate whether the following statements are true [+] or false [-] concerning the majority of project funded by the World Bank.

4.   [   ] Relatively easy to administer, measure, monitor, and to demonstrate end products.
5.   [   ] Result in substantial increases in the gross national product.
6.   [   ] Result in substantial increases in wealth for all residents of the developing country.
7.   [   ] Aggravate poverty.
8.   [   ] Require the use of modern machinery and technology.
9.   [   ] Increases diversity in crops and methods of farming.
10.   [   ] Promote nonsustainable agriculture.
11.   [   ] Increase the debt crisis in developing countries.

12.   Explain why developing nations are experiencing a population explosion and developed nations are not.

_____

Give three reasons why U.S. firms are moving their production facilities to developing nations.

13.   _____

14.   _____

15.   _____

16.   Explain how the movement of U.S. firms to developing nations has the potential of doing more harm than good for the people.

_____

_____

Indicate whether the following are [+] or are not [-] characteristics of World Bank loans applied to agriculture in developing countries.

17. [   ] cash crops   18. [   ] higher production per acre   19. [   ] use of large tractors

20. [   ] combat absolute poverty

21. The utilization of local resources without upsetting the social structure is a characteristic of projects funded by the (World, Grumman) bank?

List the three factors that are **most** directly correlated with low fertility rates.

22. _____

23. _____

24. _____

25. Family planning services do not provide

   a. counseling on reproduction and contraceptive techniques.
   b. contraceptives.
   c. unlimited abortions.
   d. counseling on the health advantages of spacing children.

CHAPTER 8

THE PRODUCTION AND DISTRIBUTION OF FOOD

Chronicles on the historical development of humankind contain countless records of how food production and distribution of food have altered the political, ecological, economic, and cultural development of societies. These chronicles demonstrate how the demise of once prominent civilizations was the direct result of inadequate food production and distribution. Note that in order for a society to maintain its agricultural integrity, it must develop and maintain a reliable agricultural production **and** distribution system. Chapter 8 contains discussions of how failure to maintain both systems results in a chaotic and functionless agricultural system. More importantly, this chapter introduces the concept of a sustainable system of producing food for the 5.7 billion people that now inhabit the earth. Chapter 7 covered the implications of an exponential increase in human population growth on food production and distribution. Can existing agricultural methods feed the worlds' human population? Are there alternative strategies that would implement and perpetuate a sustainable agricultural system?

The strategies used to increase food production in the last 40 years cannot be relied on to meet the predicted need for an increasing agricultural base in the next 40 years. We have exceeded the ecosystem carrying capacity for pesticides, fertilizers, irrigation, and tillable land. We cannot even rely on science and technology to provide any significant increases. There needs to be a new agenda in agriculture that is environmentally sensitive, equitable, relevant, and sustainable. Chapter 8 gives some insights on how this new agenda is achievable.

STUDY QUESTIONS

Crops and Animals: Major Patterns of Food Production
The Development of Modern Industrialized Agriculture

1. Indicate whether the following practices were [+] or were not [-] common agricultural practices 150 years ago.

    [　] Bringing additional land into cultivation.
    [　] Increasing use of fertilizer.
    [　] Increasing use of irrigation.
    [　] Increasing use of chemical pesticides.
    [　] Substituting new genetic varieties of grains.
    [　] Rotating crops.
    [　] Growing many different kinds of crops.
    [　] Recycling animal wastes.

The Transformation of Traditional Agriculture

2. What other revolution contributed to a revolution in agriculture? _____

3. Today, less than _____ percent of the U.S. workforce produces enough food for the nation's needs.

4. Indicate whether the following practices were [+] or were not [-] characteristics of the agricultural revolution that increased food production so dramatically in the last 40 years.

[   ] Bringing additional land into cultivation.
[   ] Increasing use of fertilizer.
[   ] Increasing use of irrigation.
[   ] Increasing use of chemical pesticides.
[   ] Substituting new genetic varieties of grains.
[   ] Rotating crops.
[   ] Growing many different kinds of crops.
[   ] Recycling animal wastes.

5. Which of the following agriculture practices will [+] and will not [-] increase food production in the **next** 40 years.

[   ] Bringing additional land into cultivation.
[   ] Increasing use of fertilizer.
[   ] Increasing use of irrigation.
[   ] Increasing use of chemical pesticides.
[   ] Substituting new genetic varieties of grains.
[   ] Rotating crops.
[   ] Growing many different kinds of crops.
[   ] Recycling animal wastes.

Machinery

6. The shift from animal labor to machinery has led to:

a. high dependency on fossil fuel energy. (True or False)
b. soil compaction. (True or False)
c. recycling of animal wastes. (True or False)

Land Under Cultivation

7. Any future attempts to bring additional land under cultivation will probably result in:

a. use of marginal, highly erodible lands. (True or False)
b. loss of forests and wetlands. (True or False)

Fertilizers and Pesticides

8. Adding more fertilizer than the optimum required leads to:

a. ever increasing productivity. (True or False)
b. increased plant vulnerability to pests. (True or False)
c. pollution. (True or False)

9. The use of pesticides in agriculture:

a. provided better pest control. (True or False)

b. increased crop yields. (True or False)
c. led to pest resistance. (True or False)
d. caused adverse side effects to human and environmental health. (True or False)

Irrigation

10.   Even though irrigated acreage increased about 2.6 times from 1950 to 1980, irrigation is increasing at a slower pace due to:

a. lack of new water sources. (True or False)
b. development of drought resistant crops. (True or False)
c. waterlogging and accumulation of salts in the soil. (True or False)

High-Yielding Varieties

11.   The new varieties of wheat and rice produced in the 1960s required more _____ and _____ but gave yields _____ to _____ when compared to traditional varieties.

12.   The high world wide production of wheat and rice resulting from the new genetic varieties was hailed as the _____ revolution.

The Green Revolution

13.   Identify which of the following aspects of the Green Revolution were benefits [B] or drawbacks [D] for the people of developing nations.

[   ] Closed the gap between food production and food needs in some countries.
[   ] The growth and production limits of "high-yielding" varieties of wheat and rice.
[   ] Heavy reliance on irrigation.
[   ] Heavy reliance on fertilizers.
[   ] Impact on farm laborers and small landowners.
[   ] Impact on culturally-specific crops.

Subsistence Agriculture in the Developing World

14.   Using the terms **MORE** or **LESS**, complete the following statements about subsistence farming. *Subsistence farming in developing countries:*

is ____more____ labor-intensive, requires ____less____ technology, utilizes ____more____ marginally productive land, and involves the clearing of ____more____ tropical forests.

Animal Farming and its Consequences

15.   What proportion of the world's crop lands are used to feed animals? _____

16.   In the United States, _____ percent of the grain crop goes to animals.

17. Indicate whether the following are [+] or are not [-] environmental consequences of animal farming.

[   ] overgrazing   [   ] deforestation of tropical rain forests
[   ] increases of atmospheric carbon dioxide

Prospects for Increasing Food Production

18. List two prospects for increasing food production in the future.

a. _____

b. _____

19. List two factors that will limit the ability to increase crop yields.

a. _____   b. _____

Food Distribution and Trade
Patterns in Food Trade

20. The world region recognized as the major source of exportable grains is _____

_____

21. Identify whether the 1990 grain import (I) export (E) ratio of each of the following countries is [a] I = E, [b] I < E, or [c] I > E

[   ] North America   [   ] Asia   [   ] Latin America   [   ] Africa   [   ] Mexico

22. For those countries that import more grain than they export, what has been the primary cause for the trade imbalance?

_____

Levels of Responsibility in Supplying Food

23. Identify the three major levels of responsibility for meeting food needs.

a. _____   b. _____   c. _____

24. Indicate whether the following policies focus on meeting the food needs of [a] family, [b] nation, or [c] globe (see Fig. 8-9).

[   ] effective safety net
[   ] food aid for famine relief
[   ] employment security
[   ] effective family planning
[   ] fair trade
[   ] just land distribution

Hunger, Malnutrition, and Famine
Extent and Consequences of Hunger

25. At least ___20___ percent of the world's people suffer from the effects of hunger and malnutrition.

26. In what order (1 to 3) are the consequences of malnutrition and hunger reflected in a population?
     ___3___ men   ___1___ children   ___2___ women

Root Cause of Hunger

27. The root cause of hunger is ___poverty___.

28. Is it (True or False) that our planet does (can) produce enough food for everyone alive today?

29. Food surpluses enter the cash economy and flow in the direction of ___economic___ demand not of ___nutritional___ need.

30. Using the answer to item #29, is it MORE or LESS probable that:

     ___more___ pet cats will be well fed.
     ___less___ undernourished children will be well fed.

Famine

31. The two primary causes of famine are _____ and _____.

32. Indicate whether the primary famine in the following countries was due to [a] drought or [b] war and the number of deaths resulting from these famines.

                                              Number of Deaths

     [   ] Sahel region of West Africa   _____
     [   ] Mozambique   _____

Food Aid

33. Indicate whether the following statements are true [+] or false [-] concerning the overall effects of providing food aid to countries in need.

     [   ] Providing food actually alleviates chronic hunger in developing countries.
     [   ] It undercuts prices for domestically produced food sold in local markets.
     [   ] The entire local economy deteriorates.
     [   ] It contributes to environmental and ecological deterioration.

34. Food aid to developing countries usually (Disrupts, Stimulates) the local economy and leads to a/an (Increase, Decrease) in local food production.

35. Food aid tends to (Improve, Aggravate) the problem of chronic hunger.

36. What was nonsustainable about the agricultural practices of the following civilizations?

    a. Sumerians: _____

    b. Greeks: _____

37. What is the goal of sustainable agriculture?

    _____

    _____

38. Which principle of ecoystem sustainability do the following agricultural practices promote?

    [   ] use of green manures
    [   ] poly cropping
    [   ] growing gardens rather than grass
    [   ] recognizing that rangelands have a carrying capacity

Final Thoughts on Hunger

39. In order to alleviate world hunger, humankind needs to develop new (a. technologies, b. forums of political and social action).

Term	Definition	Example (s)	Term	Definition	Example (s)
subsistence farming	n	u	sustainable agriculture	o	j
food security	f	e	organic farming	m	h
hunger	j	h	green manures	h	g
malnutrition	l	h	green revolution	g	f
undernutrition	p	h,x	cash crops	b	b
absolute poverty	a	a,h	exports	c	b
food aid	e	d	high-yielding	i	f
famine	d	c	imports	x	b

VOCABULARY DRILL

Directions: Match each term with the list of definitions and examples given in the tables below. Term definitions and examples might be used more than once.

VOCABULARY DEFINTIONS
a. lack of sufficient income to meet most basic biological needs
b.crops sold by a country
c. products sold by a country
d. severe shortage of food accompanied by significant increase in death rate
e. distribution of food where famine is evident
f.ability to meet the nutritional needs of all family members
g. a remarkable increase in crop production
h. crop residue
i.crops ability to produce beyond the norm through genetic manipulation
j. lack of basic food required for energy and meeting nutritional needs
k. products purchased by a country
l. lack of essential nutrients
m. regular addition of crop residues and animal manures to build up soil humus
n. farming for household needs and small cash crops
o. maintain agricultural production without degrading the environment
p. lack of adequate food energy (calories)

VOCABULARY EXAMPLES	
a. 1.2 billion people	g. lawn clippings
b. coffee, cocaine, cocoa	h. Fig. 8-14
c. Fig. 8-13	i. Fig. 8-5
d. Egypt	j. applying the four principles of ecosystem sustainability
e. the family level of responsibility	k. 786 million people
f. wheat, corn, rice	

Which of the following factors will enable humankind to increase agricultural production in the **next** 40 years (check all that apply).

1. [   ] bringing additional land into cultivation

2. [   ] increasing use of fertilizer

3. [   ] increasing use of irrigation   *all blank*

4. [   ] increasing use of chemical pesticides

5. [   ] substituting new genetic varieties of grains

Briefly explain your responses to items 1 through 5 above.

6. Bringing additional land into cultivation:_____

_____

7. Increasing use of fertilizer: _____

_____

8. Increasing use of irrigation: _____

_____

9. Increasing use of chemical pesticides: _____

_____

10. Substituting new genetic varieties of grains: _____

_____

Identify [+] those characteristics of agricultural that will be absolutely necessary to initiate future Green Revolutions in developing countries.

11.   [   ] reliance on "high-yielding" varieties of wheat and rice

12.   [+] polycropping techniques

13.   [+] application of the principles of ecosystem sustainability

14.   [+] more mobilization of the human work force in food production

15.   [+] utilization of culturally-specific crops

16.   [   ] reliance on irrigation and fertilizers

17.   There appears to be a direct relationship between increased reliance on food imports and
_____ growth in some developing countries.

Formulation of policies to meet the food needs of the global population are required at the levels of:

18. _____   19. _____   20. _____

21.   Explain how the global priorities on food distribution make it more probable that a pet cat will
be well fed than a starving child.

_____

_____

22.   Explain, from the agribusiness side of the issue, why food aid to developing countries does not
solve but rather aggravates their hunger problems.

_____

_____

CHAPTER 9

SOIL AND THE SOIL ECOSYSTEM

Chapter 8 emphasized that the underlying the support of civilization is a stable agricultural production and distribution system. No society can maintain itself for long in a civilized state without an ample reliable food supply for its citizens and such a supply can only come from the artificial propagation of various plants and animals, i.e., agriculture. Propagation of plant species is even more fundamental than that of animals because plants are always at the bottom of the food chain.

In turn, the two resources that are most fundamental in supporting plant production are fertile soil and adequate water. We take solar energy as a given since it remains essentially constant and humans can do little to manage it in one way or another. Both fertile soil and water supplies are what may be termed renewable resources since with proper management, they can be maintained, used, and reused indefinitely. But, without proper management, they can be depleted and/or destroyed. Indeed, there is much evidence to suggest the many ancient extinct civilizations came to their downfall as a result of depleting their soil and/or water resources.

Given past lessons, it would seem that we would have arrived at the point of appreciating the fundamental importance of soil and water resources, and that we would be maintaining them appropriately especially in the face our increasing population discussed in earlier chapters. Unfortunately, such is not the case. While we have the understanding and the technology to maintain soil and water resources, in too many cases we are not doing so. We are exploiting and squandering these resources in a manner that produces short-term gain but in the long-term leaves them depleted and destroyed.

Clearly, the way we are depleting and destroying water and soil resources does not hold well for future generations. Current society is not sustainable if we persist with current attitudes and methods toward depletion and destruction of these precious resources. If we wish to sustain our society we must devote considerably more thought and attention to the maintenance of soil and water resources.

STUDY QUESTIONS

1. Indicate the most probable means by which people in [a] developed and [b] developing nations obtain food.

    [   ] raise the food or gather it from natural ecosystems
    [   ] purchase the food
    [   ] have food given to them

2. List three impacts on those nations that depend heavily on food aid.

    a. _____   b. _____

    c. _____

3.   Conversion of farmland to nonfarm uses in the United States has averaged _____ acres/year over the past decade.

4.   During the last 40 years one third of the worlds' cropland has been lost to _____ and _____.

5.   List three ways cropland losses have been "replaced."

        a. _____   b. _____

        c. _____

6.   Over _____ percent of the worlds' food supply is from land-based systems that rely on _____ as the fundamental resource.

7.   The cornerstone of sustaining civilization is maintaining productive _____.

Plants and Soil

8.   List the three basic ingredients of productive topsoil (see Fig. 9-2)?

        a. _____   b. _____

        c. _____

9.   What constitutes the base of the food chain in soil ecosystems? _____

Soil for Supporting Plants

10.   The soil environment must supply the plant or at least its roots with _____, _____, and _____.

Mineral Nutrients and Nutrient-holding Capacity

11.   Three mineral nutrients required by the plant are _____, _____, and _____.

12.   The original source of mineral nutrients is from the breakdown of _____ through the process of _____.

13.   Is it (True or False) that the process of weathering is fast enough to sustain vigorous plant growth without other sources?

14.   In natural ecosystems, the greatest source of mineral nutrients for sustaining plant growth is from _____.

Water and Water-holding Capacity

15.   Is it (True or False) that plants need a more or less continuous supply of water to replace that lost in transpiration?

16. In addition to the frequency and amount of precipitation, the amount of water that is actually available to plants will depend on

    a. the amount of water that infiltrates versus running off. (True or False)
    b. the amount of water that is held in the soil versus percolating through. (True or False)
    c. the amount of water that evaporates from the soil surface. (True or False)

17. Indicate whether the following conditions need to be maximized [+] or minimized [-] in order to provide plants with the largest amount of available water.

    [ ] runoff, [ ] infiltration, [ ] water-holding capacity, [ ] surface evaporation

Aeration

18. Most plants have access to oxygen through the:

    a. roots   b. stems   c. leaves

19. List two ways of depriving plants of sufficient oxygen.

    a. _____   b. _____

Relative Acidity (pH)

20. pH is a measure of relative _____ and _____.

21. Most plants require a soil environment which is

    a. acidic, b. basic or alkaline c. close to neutral

22. Neutral is expressed by a pH of _____.

Salt and Water Uptake

23. Indicate the direction [→, ↔, ←] that water would go across the cell membrane given the following salt concentrations *outside* when compared to *inside* the plant cell wall.

    a. [   ] higher   b. [   ] lower   c. [   ] same

Soil as an Ecosystem

24. List the five ingredients required of productive topsoil.

    a. _____

    b. _____

    c. _____

    d. _____

e. _____

Soil Texture

25. **Loam** has a mineral composition of roughly ___40___ percent sand, ___40___ percent silt, and ___20___ percent clay.

26. As a soil becomes more coarse (average particle size increasing from clay to coarse sand), indicate whether each of the following soil properties will increase [+], decrease [-], or remain the same [0].

    [+] Infiltration
    [-] Surface runoff
    [-] Water-holding capacity
    [+] Aeration
    [-] Nutrient-holding capacity

27. The workability of clay soils will be (More or Less) difficult when compared to sandy soils.

Detritus, Soil Organisms, Humus, and Topsoil

28. Distinguish between **humus** [H] and **detritus** [D] in the following definitions.

    [ ]   Accumulation of dead leaves and roots on and in the soil.
    [ ]   The residue of undigested organic matter that remains after dead leaves and roots have been consumed.

29. The base (First trophic level) of the food web in soil ecosystems is

    a. producers,   b. detritus,   c. humus,   d. decomposers

30. The second trophic level of the food web in soil ecosystems is

    a. producers,   b. detritus,   c. humus,   d. decomposers

31. As soil organisms feed on the detritus and reduce it to humus, their activity also mixes and integrates humus with mineral particles developing what is called soil ___structure___.

32. The feeding and burrowing activities of organisms, which are feeding directly and indirectly on detritus

    a. break the detritus down to humus. (True or False)
    b. mix and integrate the humus with the mineral particles of the soil. (True or False)
    c. cause the soil to become loose and clumpy. (True or False)
    d. result in the formation of topsoil. (True or False)

33. In comparison to overlying topsoil, subsoil

    a. is generally darker in color. (True or False)
    b. has a loose, clumpy structure. (True or False)

34. Indicate whether the presence of humus does [+] or does not [-] improve the following soil characteristics.

   [   ] infiltration
   [   ] water-holding capacity
   [   ] nutrient-holding capacity
   [   ] aeration
   [   ] workability

35. Who gets what in the symbiotic relationship between some plants and **mycorrhizae**?

   Plants: _____

   Mycorrhizae: _____

36. Is it (True or False) that green plants support the soil organisms by being the direct or indirect source of all their food?

37. Is it (True or False) that soil organisms support green plants by making the soil more suitable for their growth?

Soil Enrichment or Mineralization

38. As humus and the clumpy soil structure break down (i.e., **mineralization**), indicate whether the following increase [+], decrease [-], or remain the same [0].

   [−] infiltration
   [−] aeration
   [+] runoff
   [−] nutrient-holding capacity
   [−] water-holding capacity
   [+] leaching of nutrients
   [−] the ability of the soil to support plants

Losing Ground

Bare Soil, Erosion, and Desertification

39. Erosion refers to soil particles being picked up by _____ or
   _____.

40. Loss of topsoil leads to

   a. increased sediment deposits in rivers. (True or False)
   b. flooding in lowlands. (True or False)
   c. a continual cycle of erosion. (True or False)

41. The erosion sequence [rank 1 to 3] from precipitation is:

   [   ] **gully**   [   ] **splash**   [   ] **sheet**

42.   Which of the following categories of soil particles would remain after wind or water erosion.

        [   ] clay   [   ] humus   [   ] silt   [   ] fine sand   [   ] coarse sand and stones

43.   Which of the following categories of soil particles are the most important soil components for nutrient and water-holding capacity?

        [   ] clay   [   ] humus   [   ] silt   [   ] fine sand   [   ] coarse sand and stones

44.   Uncontrolled wind and water erosion results in a soil ecosystem that is a functional
        _____.

45.   Is it (True or False) once destroyed, topsoil will replenish itself?

Causing and Correcting Erosion

46.   List the three major cause of soil erosion and desertification.

        a. _____, b. _____, c. _____

Overcultivation

47.   The basic reason for plowing and cultivation is for _____ control.

48.   The basic drawback to plowing is that soil is exposed to _____ and
        _____ erosion.

49.   Is it (True or False) that plowing improves aeration and infiltration?

50.   The range of erosion rates globally are _____ to _____ metric tons/hectare/year (see Table 9-4).

51.   Indicate whether the following soil characteristics increase [+] or decrease [-] with continual applications of inorganic fertilizers.

        [   ] organic matter
        [   ] nutrient content
        [   ] soil organisms
        [   ] mineralization
        [   ] desertification
        [   ] nutrient-holding capacity
        [   ] waterway pollution

52.   Is it (True or False) that chemical fertilizers is an adequate substitute for all the benefits provided by organic fertilizers?

53.   Indicate whether the following statements are true [+] or false [-] concerning inorganic (chemical) fertilizers.

        [   ] can be added to the soil efficiently and economically

[ ] are adequate substitutes for detritus
[ ] causes soil organisms to starve
[ ] enhances the mineralization process

Overgrazing

54.   Worldwide, _____ percent of rangelands suffer some degree of desertification.

55.   Overgrazing leads to

    a. less detritus to generate soil humus. (True or False)
    b. a gradual mineralization of the soil. (True or False)
    c. decreased grass production. (True or False)
    d. increased soil erosion. (True or False)
    e. decreased rainfall in the region. (True or False)

56.   Explain how BLM leases for grazing rights are a variation on the tragedy of the commons.

    _____

Deforestation

57.   A forest cover

    a. breaks the fall of raindrops reducing splash erosion. (True or False)
    b. allows water to infiltrate a litter-covered, loose topsoil. (True or False)
    c. allows 50 percent less runoff when compared to grasslands. (True or False)
    d. reduces leaching of soil nitrogen. (True or False)

58.   Which of the following is **not** a usual reason for clearing forests?

    a. To experiment with secondary succession.
    b. To permit agriculture.
    c. To obtain structural wood.
    d. To obtain firewood.

59.   Deforestation causes **MORE** or **LESS**

    a. _____ biodiversity
    b. _____ erosion
    c. _____ new seed germination
    d. _____ sediment deposits in aquatic ecosystems

60.   Tropical rainforests are being cleared at nearly _____ percent/year.

Summation

61.   The total worldwide loss of soil from crop, range, and deforested lands in _____ billion tons/year.

The Other End of the Erosion Problem

62.   Indicate whether the following conditions increase [+] or decrease [-] from the deposition of **sediments** in waterways.

[   ] flooding
[   ] secondary succession
[   ] fish kills
[   ] pollution
[   ] groundwater reserves

Irrigation, Salinization and Desertification

63.   Adding water by artificial means is called _____.

64.   Irrigated water contains at least 200-500 parts per million salt which when left behind through evaporation leads to _____.

65.   _____ percent of all irrigated land has already been salinized and an additional _____ to _____ million acres are salinized each year.

66.   One country that is becoming totally desertified is _____ (see Table 9-5).

67.   In the United States, about _____ percent of our original agricultural land has been rendered nonproductive by _____ (see Table 9-5).

"Greening" Farm Policy

68.   Making agriculture sustainable will depend upon our ability to control _____, _____, _____, and _____.

69.   Indicate whether the following descriptions of methods of controlling soil erosion apply to [a] contour farming, [b] strip cropping, [c] shelter belts, or [d] terracing (see Figure 9-18).

[   ] Planting rows of trees around fields.
[   ] Plowing and cultivating at right angles to the slope.
[   ] Planting alternative strips of grass between strips of corn.
[   ] Grading slopes into a series of steps.

70.   List the four objectives of sustainable agriculture.

a. _____

b. _____

c. _____

d. _____

71. Indicate whether the following conservation measures were part of the [a] Conservation Reserve Program of 1985, [b] the Food Security Act of 1985, or [c] 1990 Farm Bill

[   ] Farmers were paid to put highly erodible land into a Conservation Reserve program.
[   ] Farmers were required to develop and implement a soil conservation program.
[   ] Encourages farmers to save wetlands.
[   ] Resulted from the lobbying efforts of the **urban environmental coalition**.

72. Which of the following practices meet the objectives of sustainable agriculture? (Check all that apply)

[   ] use of organic fertilizers
[   ] use of inorganic fertilizers exclusively
[   ] crop rotation
[   ] strip cropping
[   ] drip irrigation
[   ] dryland farming
[   ] use of chemical pesticides
[   ] biological control of pests
[   ] monoculture farming

73. Relate the following practices with [T] traditional or [A] alternative farming practices.

[   ] feed lots   [   ] polycropping   [   ] monocultures   [   ] lower $ inputs

74. Past national farm policies promoted (Traditional, Alternative) farming practices.

75. Present national farm policies promote (Traditional, Alternative) farming practices.

Directions: Match each term with the list of definitions and examples given in the tables below. Term definitions and examples might be used more than once.

Term	Definition	Example (s)	Term	Definition	Example (s)
compaction	n	b	organic fertilizer	f	c, j
composting	t	x	overcultivation	ee	s
deforestation	aa	o	overgrazing	ff	t
desertification	bb	p	pH	o	h
erosion	ii	q	salinization	gg	u
evaporative water	l	d	sediments	hh	v
fertilizer	e	c	soil structure	u	l
gully erosion	cc	q	soil texture	p	i
humus	s	j	soil aeration	m	f
infiltrate	k	d	soil profile	x	l
inorganic fertilizer	b	c	soil fertility	a	a
irrigation	j	e	sheet erosion	jj	q
loam	q	i	splash erosion	kk	q
mineral nutrients	b	b	subsoil	w	l
mineralization	z	n	topsoil	v	l
mycorrhizae	y	m	transpiration	i	d
no-till agriculture	dd	r	water-holding capacity	h	d
nutrient-holding capacity	c	c	weathering	d	c
			workability	r	i

VOCABULARY DEFINITIONS

a. soil's ability to support plant growth

b. nutrients present in rocks

c. capacity to bind and hold nutrient ions

d. gradual chemical-physical breakdown of rock

e. material that contains one or more necessary nutrients

f. organic material that contains one or more necessary nutrients

g. inorganic material that contain one or more necessary nutrients

h. soil's ability to hold water after it infiltrates

i. transpiration: loss of water through plant leaves

j. artificial means of providing water to croplands

k. soil's ability to allow water to soak in

l. water loss from the soil surface

m. soil's ability to allow diffusion of oxygen and carbon dioxide

n. packing of soil

o. measure of hydrogen ion concentration in solution

p. relative proportions of sand, silt, and clay in soil

q. a specified mixture of sand, silt, and clay

r. workability: the ease with which soil can be cultivated

s. residue of partially decomposed organic matter

t. process of fostering decay of organic wastes under more-or-less controlled conditions

u. a loose clumpy characteristic of soil

v. a dark humus-rich soil

w. a light brown, humus-poor soil

x. a layering of top and subsoil

y. soil organisms in a symbiotic relationship with plant roots

z. decomposition or oxidation of humus

aa. process of increasing soil loss through clear-cut logging

bb. effect of topsoil loss

cc. erosion when runoff water forms rivulets and streams

VOCABULARY DEFINITIONS
dd. process of reducing soil cultivation
ee. process of increasing soil loss through cultivation
ff. process of increasing soil loss through grazing too many livestock
gg. intolerable increase in soil salinity
hh. soil particles released from topsoil during erosion
ii. the process of soil and humus being picked up and carried away by wind or water
jj. water erosion resulting from lack of infiltration
kk. impact of falling raindrops

VOCABULARY EXAMPLES	
a. optimum amounts of mineral nutrients, water, oxygen	l. Fig. 9-9
b. phosphate, potassium, calcium	m. fungi
c. Fig. 9-3	n. Fig. 9-12
d. Fig. 9-5	o. Fig. 9-20
e. Fig. 9-22	p. Fig. 9-15
f. oxygen concentration	q. Fig. 9-13
g. excessive foot or vehicular traffic	r. Fig. 9-17
h. $0 \leftrightarrow 7 \leftrightarrow 14$	s. plowing, discing, harrowing
i. 40% sand, 40% silt, 20% clay	t. BLM lands
j. Fig. 9-8	u. Fig. 9-23
k. making humus	v. muddy river

1. List the organisms in the soil ecosystem that would appear at each of the trophic levels indicated in the box below.

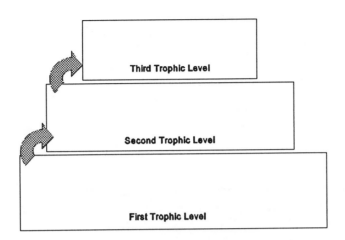

Complete the table below by indicating the relationship of the factors listed with the soil properties given in the table. The relationship can be designated as **excellent [E], good [G], medium [M], or poor [P].**

FACTORS

Soil Property	Water Infiltration	Water-holding Capacity	Nutrient-holding Capacity	Aeration
Sand	2. G	3. P	4. P	5. G
Silt	6. M	7. M	8. M	9. M
Clay	10. P	11. G	12. G	13. P
Loam	14. M	15. M	16. M	17. M
Humus rich	18. E	19. E	20. E	21. E

Explain how the practice of inefficient wood burning and scouring the landscape of leaves, sticks, manure, and other forms of detritus have a negative effect on the structure and function of the soil ecosystem.

22. Negative effect on STRUCTURE: the detritus struc. is lost which forms the base of the soil ecosystem food chain and one of the primary soil erosion preventatives.

23. Negative effect on FUNCTION: _Wout detritus for detritus feeders and decomposers, transfer of energy and nutrients to other soil biota cannot take place_

Which of the agricultural practices below will increase [+] and decrease [-] soil erosion?

24.   [+] plowing

25.   [+] using inorganic fertilizers exclusively

26.   [+] overgrazing

27.   [+] deforestation

28.   [−] planting shelter belts

29.   [−] strip cropping

Complete the below table by first identifying those agricultural practices that do [+] and do not [-] meet the objectives of sustainable agriculture and second whether these practices are traditional [T] or alternative [A] types.

Agricultural Practices	Sustainable	Type
Use of organic fertilizers	30. [+]	31. [A]
Center pivot irrigation	32. [−]	33. [T]
Use of chemical pesticides	34. [−]	35. [T]
Dryland farming	36. [+]	37. [A]
Crop rotation	38. [+]	39. [A]

CHAPTER 10

PESTS AND PEST CONTROL

The struggle between man and insects began long before the dawn of civilization, has continued without cessation to the present time, and will continue, no doubt, as long as the human race endures. We commonly think of ourselves as the lords and conquerors of nature. But insects had thoroughly mastered the world and taken full possession of it before man began the attempt. They had, consequently, all the advantages of possession of the field when the contest began, and they have disputed every step of our invasion of their original domain so persistently and successfully that we can even yet scarcely flatter ourselves that we have gained any very important advantage over them. If they want our crops, they still help themselves to them. If they wish the blood of our domestic animals, they pump it out of the veins of our cattle and our horses at their leisure and under our very eyes. If they choose to take up their abode with us, we cannot wholly keep them out of the houses we live in. We cannot even protect our very persons from their annoying and pestiferous attacks, and since the world began, we have never yet exterminated---we probably shall never exterminate---so much as a single insect species.

(THE BUGS ARE COMING, Time Magazine, July 12, 1976, page 38.)

When someone says the word "pest" to you, what is the first image that comes to mind? For most people, the image is a bug! This chapter will broaden you understanding of the sorts of organisms that are considered pests in the human ecosystem. In fact, you will see how the structure and function of the human ecosystem is predestined to be plagued by all sorts of pests. The very definition of a pest, i.e., "any organism that impacts negatively on human health or economics," should help you to visualize a whole host of organisms that would fit this definition. For example, organisms that cause disease, harass us, feed on our crops and domestic livestock, cause the deterioration of our goods and food, or detract from our quality of life in any way are **pests**. Pest organisms can be plants, animals, bacteria, and parasites among others.

It is an easy task to identify the pest. It is much more difficult to figure the best procedures for eradicating it or, more importantly, how to control a pest and maintain the integrity of the ecosystem. Chapter 10 provides a historical overview of our early attempts at pest control that did not take the integrity of the ecosystem into consideration. Chapter content demonstrates how quick and dirty solutions to a complex problem can back fire. It is important to emphasize that the early attempts at pest control were complete failures and left the pests in a more virulent condition. Additionally, the natural ecosystem was inundated with vast quantities and types of chemicals (pesticides) which had the most adverse effects on non-pest organisms. Alternative pest control methods are available, effective in pest control, and examples of applications of the principles of ecosystem sustainability.

STUDY QUESTIONS

1.   Is it (True or False) that the term **pest** refers only to bugs?

The Need for Pest Control

Match the organisms on the left with their pest category on the right.

<u>Organism</u>   <u>Pest Category</u>

2.   [   ] weeds   a. disease
3.   [   ] molds   b. annoyance
4.   [   ] coyotes   c. competitive feeder
5.   [   ] snails   d. predator
6.   [   ] bacteria   e. decomposer
7.   [   ] mosquitos   f. nutrient competitor

The Importance of Pest Control

8.   Is it (True or False) that humankind need no longer rely on pest control.

Crop Losses Due to Pests

9.   An estimated _____ percent of agricultural production are lost to pests now compared to _____ percent in 1950s.

10.   Indicate whether the following chemicals are meant to target (a. plants or b. animals).

   [   ] pesticides   [   ] herbicides

Different Philosophies of Pest Control

11.   List the two philosophies of pest control.

   a. _____   b. _____

12.   Which philosophy treats only the symptoms of pest outbreaks? _____

13.   Which philosophy stimulates ecosystem immunity to pest outbreaks? _____

14.   Which philosophy combines the approaches of 11a and b? _____

Promises and Problems of the Chemical Approach

Development of Chemical Pesticides and Their Successes

15.   Indicate whether the following chemicals are meant to target: [a] all organisms, [b] mice and rats, [c] insects, or [d] fungi.

   [   ] fungicides   [   ] rodenticides   [   ] insecticides   [   ] biocides

16. A common chemical pesticide used in the 1930s was

    a. cyanide,   b. arsenic,   c. sulfur,   d. DDT

17. Indicate whether the following are characteristics of [a] first-generation, [b] second-generation, or [c] both types of pesticides.

    [   ] Consisted of toxic heavy metals.
    [   ] Are inorganic compounds.
    [   ] Are synthetic organic compounds.
    [   ] Are persistent.
    [   ] Promote pest resistance.
    [   ] Are broad spectrum in their effect.
    [   ] Examples are cyanide and arsenic.
    [   ] An example is DDT (dichlorodiphenyltrichloroethane)
    [   ] Stimulated Rachel Carson to write *Silent Spring*

Problems Stemming From Chemical Pesticide Use

18. List three problems associated with the use of synthetic organic pesticides.

    a. _____

    b. _____

    c. _____

Development of Pest Resistance

19. Chemical pesticides gradually lose their _____.

20. Loss of effectiveness leads to two choices by growers, namely, use (More, Less) or use the (Same, New) pesticides.

21. The use of chemical pesticides on insects leads to selection for the (Sensitivity, Resistance) allele in the gene pool of the population.

22. Each generation of insects exposed to a chemical pesticide will have (More, Less) alleles for resistance in the gene pool of the population.

23. Is it (True or False) that as a pest population becomes resistant to one pesticide it may, at the same time, become resistant to others to which it has not yet been exposed?

Resurgences and Secondary Pest Outbreaks

24. Match the following definitions with [a] resurgence or [b] secondary pest outbreak.

    [   ] Insects that were previously not a problem, become a pest category.
    [   ] The return of the pest at higher and more severe levels.

25.   Pesticide treatments often have a (Less, Greater) effect on the natural enemies of pests.

26.   Which of the following statements are [+] or are not [-] explanations of why pesticides have a greater impact on natural enemies than upon the target pest?

    [   ] Herbivore species are intrinsically more resistant to the pesticides than are its predators.
    [   ] Biomagnification gives the predator a higher dose of the pesticide.
    [   ] Predatory species may be starved out due to a temporary lack of prey.

27.   How is the chemical approach contrary to basic ecological principles?

_____

_____

Adverse Environmental and Human Health Effects

28.   What were three sources of evidence that indicated DDT was the causative factor in reproductive failure in fish-eating birds?

    a. DDT levels in fragile egg shells were (High, Low).
    b. DDT (Promotes, Reduces) calcium metabolism.
    c. DDT levels in fish-eating birds was (Higher, Lower).

29.   The highest DDT levels can be expected at the (Bottom, Top) of the food chain.

30.   DDT accumulates in the body _____ of humans and other animals.

31.   DDT was banned in the United States in the early 19_____s.

32.   Is it (True or False) that when DDT was banned, pesticides were no longer used?

33.   About _____ million tons of pesticides are used globally.

34.   If less than one percent of the pesticides dumped on the environment come into contact with the pest organism, where does the remaining 99 percent of the pesticide go?

    a. Drifts into the air? (True or False)
    b. Settles on surrounding ecosystems? (True or False)
    c. Settles on bodies of water? (True or False)
    d. Remains as residue on foods? (True or False)
    e. Leaches from the soil into aquifers? (True or False)
    f. Is captured and recycled by growers? (True or False)

The Pesticide Treadmill

35.   Construct the last four steps in the pesticide treadmill below using [a] new and larger quantities of chemicals and [b] more resistance and secondary outbreaks.

    [a] Step 1 → [b] Step 2 → [   ] Step 3 → [   ] Step 4 → [   ] Step 5 → [   ] Step 6 → ?

Nonpersistent Pesticides: Are They the Answer?

36.   Assume that 1000 pounds of DDT was applied to a field in Nebraska in 1950. How much of the 1000 pounds will be left in:

a. 1970 _____, b. 1990 _____, c. 2010 _____

d. When will the DDT disappear? _____

37.   The half-life of nonpersistent pesticides is (a. days, b. weeks, c. years, d. a or b).

38.   Indicate whether the following statements are true [+] or false [-] concerning nonpersistent pesticides.

[   ] Stay in the environment for a long time.
[   ] Are less toxic than DDT.
[   ] Are broad spectrum pesticides.
[   ] Do not cause resistance in pest species.
[   ] Do not cause resurgence or secondary outbreaks.

Alternative Pest Control Methods

39.   List the four alternative pest control methods.

a. _____

b. _____

c. _____

d. _____

Cultural Controls

Cultural Control of Pests Affecting Humans

40.   Indicate whether the following control measures are in the category of [a] sanitation or [b] personal hygiene.

[   ] water purification
[   ] bathing
[   ] wearing clean clothing
[   ] proper and systematic disposal of garbage
[   ] using clean cooking and eating utensils
[   ] refrigeration, freezing, and canning foods

Cultural Control of Pests Affecting Lawns, Gardens, and Crops

41. Indicate whether the following control measures are examples of

   a. Selection of what to grow and where to grow it
   b. Management of lawns and pastures
   c. Water and fertilizer management
   d. Timing of planting
   e. Destruction of crop residues
   f. Bordering crops with beneficial "weeds"
   g. Crop rotation
   h. Polyculture
   i. Quarantines

   [　] Prohibiting biological materials that carry pests from entering a country.
   [　] Planting plants under their optimum conditions.
   [　] Growing two or more species together or in alternate rows.
   [　] Cutting grass no less than 3 inches.
   [　] Alternate crops from one year to the next in a given field.
   [　] Providing the ideal levels of water and fertilizer.
   [　] Plowing under or burning the residue left after harvest.
   [　] Eliminate plants that act as pest attractants and grow others that act as repellents.
   [　] Delay planting so most of the pest population starves before the plants are available.

Control by Natural Enemies

42. Match the following examples of pests with the natural enemy that was used to control each pest (see Figures 10-10 and 10-11)

Pest	Natural Enemy
[　] scale insects	a. manatees
[　] caterpillars	b. plant-eating insects
[　] gypsy moths and Japanese beetles	c. bacteria
[　] prickly pear cactus	d. parasitic wasps
[　] water hyacinths	e. vedalia (ladybird) beetles
[　] rabbits in Australia	

43. The first step in using natural enemies is _____ of the natural enemies that already exist.

44. Of the 2000 serious insect pests, _____ percent remain without effective natural enemies.

45. Is it (True or False) that finding natural enemies to pests generates as much profit for industry as do synthetic chemicals?

46. Natural enemies are a (Short, Long) term form of pest control.

47. Natural enemies are (More, Less) profitable in the long term than synthetic chemicals.

Genetic Control

48. Is it (True or False) that plant-eating insects have a wide variety of host plants?

49. There is a genetic (Compatibility, Incompatibility) between the pest and species that are not attacked.

50. The potential of _____ control is implied by genetic incompatibility between the pest and species that are not attacked.

Chemical Barriers

51. A chemical barrier means that the plant produces some chemical that is _____ or _____ to potential pests.

52. A chemical barrier was introduced into wheat plants that killed Hessian fly _____ .

Physical Barriers

53. Physical barriers are _____ traits that impede pest attacks.

54. An example of a physical barrier is _____ hairs on leaf surfaces.

55. Is it (True or False) that pests cannot overcome genetic controls through resistance?

Control With Sterile Males

56. The sterile male procedure involves subjecting the (a. egg, b. larvae, c. pupa, d. adults) to just enough high energy radiation to render them sterile.

57. Is it (True or False) that pests cannot develop resistance to the sterile male technique?

58. An example of application of the sterile male technique is the _____ fly.

Natural Chemical Control

59. Indicate whether the following definitions pertain to [a] hormones or [b] pheromones.

    [   ] A chemical secreted by one insect that influences the behavior of another.
    [   ] Chemicals that provide signals for development and metabolic functions.

60. The four aims of natural chemical control are to _____, _____, _____, and _____ an insect's own hormones or pheromones to disrupt its life cycle.

61. The two advantages of natural chemicals are that they are _____ and _____ .

62. Juvenile hormone had a direct adverse affect on the (a. egg, b. larva, c. pupa, d. adult).

63. Sex attractant pheromones may be used in the _____ or _____ techniques.

64. Is it (True or False) that the potential of future control of pests through natural chemicals is promising?

Socioeconomic Issues In Pest Management

Pressures to Use Pesticides

65. Natural controls are aimed at (Management, Eradication) of pest populations.

66. Significant damage is implied when the cost of damage is considerably (Greater, Less) than the cost of pesticide application.

67. Indicate whether the following attitudes or actions are examples of:

a. cosmetic spraying.   b. insurance spraying.   c. chemical company vested interest.

[   ] Consumer desire for unblemished produce.
[   ] The use of pesticides "just to be safe."
[   ] Ignorance of whether a "bug" is causing damage or beneficial.
[   ] The "Snow White" syndrome
[   ] Promoting and exploiting the above attitudes.

Organically Grown Food

68. Control of pests in organic farming is likely to consist of (Check all that apply):

[   ] nonpersistent pesticides
[   ] natural controls
[   ] cultural controls
[   ] synthetic organic pesticides

Integrated Pest Management

69. Integrated pest management (IPM) addresses all _____, _____, and _____ factors.

70. IPM is (Short, Long) term approach.

71. IPM (Never, Sometimes) requires use of chemical pesticides.

Public Policy

FIFRA and Its Problems

72. What is the law that regulates pesticide use? _____

73. The law requires that manufacturers _____ pesticides with the government before marketing them.

74. The registration procedure requires _____ to determine toxicity to animals.

75. List four shortcomings of FIFRA.

a. _____

b. _____

c. _____

d. _____

76. Indicate whether the following statements reflect FIFRA shortcomings related to [a] inadequate testing, [b] bans on a case-by-case basis, [c] pesticide exports, or [d] lack of public trust.

[   ] 1400 chemicals and 40,000 formulations.
[   ] 100 million pounds of pesticides shipped to less developed countries.
[   ] EDB found in hundreds of wells in Florida and California.
[   ] Using 100 to 200 mice over a one- to two-year period.
[   ] Substituting healthy mice for experimental mice that died.
[   ] Intense lobbying efforts by the chemical industry as opposed to public forums.

The Delaney Clause

77. The Delaney Clause links pesticide _____ in food with risk of _____.

78. The two sides of the argument concerning the Delaney Clause focus on _____ and _____.

Pesticides in Developing Countries

79. What is the annual export of pesticides to developing countries by the United States? _____ metric tons

80. Is it (True or False) that these pesticide exports do not include those banned in the U.S.?

81. Explain how PIC and FAO Code of Conduct are protecting developing countries from imports of unwanted pesticides.

PIC: _____

FAO: _____

New Policy Needs

82.   What new pesticide regulatory rules are needed to offset the shortcomings of FIFRA? (Check all that apply).

[   ] groundwater protection
[   ] food safety
[   ] measuring extent and dosage of pesticide exposures
[   ] worker protection
[   ] expanded pesticide testing requirements

83.   List two pesticide legislative reforms that are being considered.

a. _____

b. _____

VOCABULARY DRILL

Directions: Match each term with the list of definitions and examples given in the tables below. Term definitions and examples might be used more than once.

Term	Definition	Example (s)	Term	Definition	Example (s)
biomagnification			natural chemicals		
broad-spectrum			nonpersistent		
chemical barriers			organically grown		
cosmetic spraying			persistent		
cultural control			pest		
economic threshold			pesticide treadmill		
FIFRA			pesticide		
first-generation			physical barriers		
genetic control			resistance		
herbicide			resurgence		
insect life cycle			second-generation		
insurance spraying			secondary outbreaks		
IPM			sex attractant		
natural enemies					

VOCABULARY DEFINITIONS

a. any organism that has a negative impact on human health or economics

b. chemicals that kill animal pests

c. chemicals that kill plants

d. pesticides used before the 1930s

e. pesticides used during and after the 1930s

f. pesticides that do not select target species only

g. pesticides that stay in the ecosystem for a long time

h. pest organism's ability to develop genetic immunity to a pesticide

i. pest population explodes to higher and more severe levels

j. population explosions of pests that previously were of no concern

k. the process of accumulating higher and higher doses through the food chain

l. cycle of pesticide use, resistance, resurgence, pesticide use, etc.

m. pesticides that remain for only a short period of time in the ecosystem

n. when economic losses due to pest damage outweigh cost of applying pesticide

o. use of pesticides to control pests that harm only item's outward appearance

p. use of pesticides when no control need is evident

q. metamorphic stages in an insect's life

r. chemical substance the alters the reproductive behavior of the opposite sex of the same species

s. identifying and using the natural predator of the pest species

t. developing a genetic incompatibility between the pest and its host species

u. alteration of one or more environmental factors that attract pest species

v. alteration of one or more stages in a pest species' life cycle

w. substances that are lethal or repulsive to would-be pests

x. structural traits that imped the attack of a pest

y. involves the integration of several sequential pest control techniques

y. key legislation to control pesticides in the United States

z. crop production without the use of synthetic chemical pesticides or fertilizers

VOCABULARY EXAMPLES	
a. flies, tick, mosquitos	m. "just to be safe"
b. DDT	n. Fig. 10-8
c. Round-up	o. pheromones
d. arsenic, cyanide	p. Fig. 10-10
e. Fig 10-3	q. Fig. 10-12
f. all pest species	r. Fig. 10-9
g. cotton mites	s. skunk odor
h. 0.001 → 0.01 → 0.10 → 1.0 ppm	t. hairy leaves
i. Fig. 10-7	u. Federal Insecticide, Fungicide, and Rodenticide Act
j. parathion	v. Integrated Pest Management
k. Fig. 10-14	w. Fig. 10-15
l. grocery store apples	

SELF-TEST

1. Define the term **pest**:

    _____

    Identify whether the following characteristics pertain to [a] first generation, [b] second generation, [c] both first and second generation, [d] nonpersistent, or [e] all pesticides.

2. [   ] Are broad spectrum in their effect.

3. [   ] Are harmful to humans and the environment.

4. [   ] Cause pest resistance.

5. [   ] Cause resurgence.

6. [   ] Cause secondary outbreaks.

    If a pesticide has been present in an aquatic ecosystem for several years, match the pesticide levels given below with various parts of an aquatic ecosystem.

    Pesticide Levels: a. 0.02 ppm, b. 2000 ppm, c. 40-300 ppm, d. 5 ppm, e. 1600 ppm

7. [   ] water   8. [   ] fish-eating birds   9. [   ] algae   10. [   ] plant-eating fish

11. Give the term that is used to explain your answers to 7 through 10:_____

   What are three sources of evidence that implicated DDT as the causative factor in reproductive failure in fish-eating birds?

12. _____

13. _____

14. _____

15. Explain the association between **resistance** and **resurgence**.

   _____

   _____

   List four ways of controlling pest species that do not include the use of inorganic or synthetic organic chemicals.

16. _____   17. _____

18. _____   19. _____

   Match the following actions with one of the ways (16-19) of controlling pests that you listed above.

20. _____ using high energy radiation   21. _____ crop rotation

22. _____ use of pheromone   23. _____ using ladybug beetles

24. _____ chemical barriers   25. _____ disposal of garbage

26. _____ juvenile hormone   27. _____ destruction of crop residues

28. Ecological pest management seeks to control pests by

   a. manipulating natural factors affecting the host and pest relationship.
   b. letting nature take its course.
   c. using pesticides that are known to be ecologically sound.
   d. using biological controls only.

29. Which of the following shortcomings in FIFRA promotes heavy lobbying by the chemical industry?

   a. Inadequate testing.
   b. Lack of public input.
   c. Pesticide exports.
   d. Ban issued on a case-by-case basis when threats are proven.

30.   Cosmetic spraying is done to

       a. improve the appearance of produce.
       b. improve the taste.
       c. improve the storability.
       d. protect the public health.

CHAPTER 11

WATER, THE WATER CYCLE, AND WATER MANAGEMENT

Just outside Phoenix, which gets less than 8 inches of rain a year, the fountains shoot high into the arid air, gurgling defiance of the laws of evaporation, advertising the subdivisions whose green lawns eat further and further into the Arizona desert. But in parts of the Texas High Plains, the water that made farms green with cotton and melons is giving out; land is now brown with tumbleweeds, which farsighted agronomists are investigating as a cash crop. And in stately Greenwich, Conn., surely one of the last places from which civilization will vanish, suburban matrons guard water like Bedouins, and town officials lay plans for slit-trench latrines against the not-too-distant day when the reservoirs may run dry. (Newsweek, Are We Running Out of Water? February 23, 1981, page 26)

No life can exist without water. Water is fundamental in the support of all ecosystems. Water is fundamental in the production of all food crops and animals. We use water to flush away wastes. In cooling we often use water to carry away waste heat. We also require water for drinking, recreation, and its aesthetic charm in fountains and falls. It would be hard to argue that any resource is more precious than water. Yet, we are squandering and depleting our water resources. We are making water unusable and unable to support desired ecosystems by our pollution.

There is little question that our apathy toward, our lack of understanding of, and our blatant disregard for water resources is placing all life on Earth in jeopardy. We desperately need to understand the nature of water and our water resources. More importantly, we need to use this understanding with care and wisdom to conserve and care for our water resources. Our lives will depend on it.

STUDY QUESTIONS

1. Water covers _____ percent of the Earth's surface.

2. _____ percent of the earth's water is saltwater.

3. Eighty-seven percent of the fresh water is bound up in:

    a. _____   b. _____ and c. _____

4. Only _____ percent is accessible fresh water.

5. It is possible that nations may fight wars over _____ .

The Water Cycle

6. The earth's **water cycle** is also called the _____ cycle.

7. Water enters the atmosphere through _____ and _____ .

8. Water returns from the atmosphere through _____ and _____ .

Evaporation, Condensation, and Purification

9.   The three physical states of water are _____, _____, and _____.

10.   These three states are the result of different degrees of _____ bonding between water molecules.

11.   What are the two forces that determine the different degrees of interaction between water molecules?

      a. _____   b. _____

12.   Hydrogen bonding tends to (Attract, Separate) water molecules.

13.   Kinetic energy tends to (Attract, Separate) water molecules.

14.   Below freezing, (Hydrogen Bonding, Kinetic Energy) is the dominant force determining the association between water molecules together.

15.   Above freezing, (Hydrogen Bonding, Kinetic Energy) is the dominant force determining the association between water molecules.

16.   During **evaporation**, kinetic energy is (High, Low) when compared to hydrogen bonding.

17.   During **condensation**, kinetic energy is (High, Low) when compared to hydrogen bonding.

18.   Evaporation and condensation are two water purification processes similar to _____.

19.   Is it (True or False) that all the natural freshwater on earth comes from this distillation process?

20.   Water vapor enters the atmosphere by way of:

      a. _____ from all water and moist surfaces.
      b. _____ which is really evaporation through the leaves of plants.

21.   The amount of water vapor held in the air at various temperatures is called _____ humidity.

22.   Warm air holds (More or Less) water than cold air.

Precipitation

23.   Rising air results in (High, Low) precipitation whereas descending air results in (High, Low) precipitation.

24. Air rises over equatorial regions and descends over subequatorial regions (see Figure 11-5), therefore

a. rainfall is (High, Low) in equatorial regions.
b. rainfall is (High, Low) in subequatorial regions.

25. The **rainshadow** (see Figure 11-6) refers to a region of (High, Low) precipitation.

Water Over and Through the Ground

26. When precipitation hits the ground, what two alternative pathways might it follow?

a. _____   b. _____

27. All ponds, lakes, streams, and rivers are referred to as _____ waters.

28. Water that infiltrates may follow either of two pathways including:

a. _____ or b. _____ water.

29. Plants draw mainly from _____ water.

30. Water that percolates down through the soil eventually comes to an _____ layer and accumulates filling all the empty pores and spaces. This accumulated water is now called _____ water and its upper surface is called the _____ table.

31. The layers of porous material through which groundwater moves are called an _____.

32. Where water actually enters an aquifer is called the _____ area.

33. Natural exits of groundwater to the surface are _____ or _____.

34. Springs and seeps feed streams and rivers becoming part of _____ water.

Summary of the Water Cycle

35. Summarize the below events in the water cycle by indicating whether they primarily occur as part of the [a] surface runoff loop, [b] evaporation-transpiration loop, or [c] groundwater loop (see Figure 11-3).

[ ] springs   [ ] water vapor
[ ] water table   [ ] capillary water
[ ] transpiration   [ ] gravitational water
[ ] percolation   [ ] aquifers
[ ] surface runoff   [ ] ground water
[ ] cloud formation   [ ] infiltration

36. The two filters for water that infiltrates through the ground are _____ and _____ .

37. Match the terms on the left with the water descriptions on the right (see Table 11-1).

Water Terms   Water Descriptions

[   ] Fresh   a. a mixture of fresh and salt water
[   ] Salt   b. salt concentration < 0.01%
[   ] Brackish   c. salt concentration at least 3%
[   ] Hard   d. high mineral content
[   ] Soft   e. low mineral content

Human Impacts on the Water Cycle

38. List the three main impacts that humans have on the water cycle.

   a. _____

   b. _____

   c. _____

Sources and Uses of Fresh Water

39. The two primary human concerns regarding fresh water are _____ and _____ water.

40. What proportion of the world's water resources are used for:

   a. irrigation _____%   b. industry _____%
   c. human consumption _____%

41. The major source of fresh water are: a. ground water.   b. surface water.

42. Center pivot irrigation systems use huge amounts of (Surface, Ground) water and may use as much as (1, 10, 100) thousand gallons per minute.

Overdrawing Water Resources

Consequences of Overdrawing Surface Waters

43. No more than _____ percent of a river's average flow can be taken risking shortfalls.

44. However, the demand on some rivers exceeds _____ percent of the average flow.

45. What are the consequences of overdrawing surface waters?

    a. wetlands along and at the mouth of the river drying up. (True or False)
    b. estuaries at the mouth of the river becoming increasingly salty. (True or False)
    c. dieoffs of large populations of wildlife. (True or False)

Consequences of Overdrawing Groundwater

46. Depletion of ground water will very likely result in

    a. cutbacks in agricultural production. (True or False)
    b. diminishing surface water. (True or False)
    c. adverse effects on stream and river ecosystems. (True or False)
    d. land subsidence. (True or False)
    e. formation of sinkholes. (True or False)
    f. saltwater intrusion into wells in coastal regions. (True or False)
    g. required cutbacks in municipal consumption. (True or False)

Obtaining More Water

47. It may be possible to increase water supplies:

    a. at modest cost. (True, False)
    b. without severe ecological impacts. (True, False)

48. Future water shortages will

    a. become increasingly common. (True or False)
    b. be of longer duration. (True or False)
    c. affect increasingly large areas and numbers of people. (True or False)
    d. go away. (True or False)

Using Less Water

49. The greatest demands on U.S. fresh water (Consumptive, Nonconsumptive) uses particularly in (see Table 11-2):

    a. agriculture,   b. electrical power production,   c. industry   d. homes

50. About _____ percent of irrigation water is wasted through evaporation.

51. A water conserving alternative to surface and sprinkling irrigation is _____ irrigation.

52. Water consumption in modern homes averages around _____ gallons per person per day.

53. The most effective conservation technique would be to use (More, The Same, Less) water.

54. An example of recycling water is _____ water.

Desalting Sea Water

55.   What are the two technologies used for desalinization:

       a. _____   b. _____

56.   What is the cost per 1,000 gallons of desalinized water? $_____

57.   Farmers currently pay $_____ per 1,000 gallons of fresh water for irrigation.

Storm water Mismanagement and Management

Flooding
Streambank Erosion

58.   As land is developed, natural soil surfaces are replaced by various hard surfaces. By this action

       a. infiltration is (Increased or Decreased).
       b. runoff is (Increased or Decreased).

59.   As a result of the change in infiltration and runoff indicated above predict whether each of the
       following will increase [+], decrease [-], or remain the same [0].

       [   ] amount of groundwater recharge   [   ] sediment deposition in the stream channel
       [   ] level of water table   [   ] breadth of stream channel
       [   ] amount of flow from springs   [   ] depth of stream channel
       [   ] stream flow between rains   [   ] pollution of stream channel
       [   ] stream flow during rains   [   ] frequency of flooding
       [   ] stream bank erosion   [   ] heights of floods

Increased Pollution

60.   List the categories of pollutants that enter streams and rivers from surface runoff.

       a. _____   b. _____   c. _____

       d. _____   e. _____   f. _____

       g. _____

Channelization

61.   To alleviate the problem of urban flooding , streams may be **channelized**. In this process

       a. a channel is dug out. (True or False)
       b. curves are smoothed out. (True or False)
       c. the channel is lined with concrete or rock. (True or False)
       d. a natural water flow is restored. (True or False)
       e. a natural stream ecology is restored. (True or False)

Improving Storm Water Management

62. Describe one technique for improving storm water management (see Fig. 11-26).

_____

Groping for a Better Way

63. Current trends in water use (Are, Are Not) sustainable.

64. List the four stakeholders in future water use decisions.

a. _____   b. _____

c. _____   d. _____

65. What seems to be a workable strategy for mitigating conflicting water use needs of stakeholder groups?

_____

VOCABULARY DRILL

Directions: Match each term with the list of definitions and examples given in the tables below. Term definitions and examples might be used more than once.

Term	Definition	Example (s)	Term	Definition	Example (s)
aquifer	aa	n, q	polluted water	k	h
capillary water	p	m	precipitation	t	n, o
channelization	jj	aa	rain shadow	m	j
condensation	e	e	recharge area	z	s
consumptive water use	cc	t	relative humidity	i	f
desalinization	ll	cc	saltwater	j	g
drip irrigation	hh	y	saltwater intrusion	ff	w
evaporation	c	c, w	sinkhole	dd	u
evapotranspiration	s	n	spring	bb	u
fresh water	f	b	Storm water management	un	dd
gravitational water	w	n, r	surface waters	o	l
gray water	ii	z	surface runoff	u	n, p
groundwater	v	n, q	transpiration	mm	n
humidity	h	c	water purification	l	i
hydrological cycle	r	n	water cycle	r	n
infiltration	q	n	water table	y	n, q
land subsidence	ee	v	water vapor	g	c
nonconsumptive water use	gg	x	water quality	b	b
percolation	x	n	water quantity	a	a
physical states of water	d	d	watershed	w	x
			xeroscaping	kk	bb

120

VOCABULARY DEFINITIONS
a. amount of fresh water
b. purity of fresh water
c. water molecules entering the atmosphere
d. water conditions given different temperatures
e. water molecules rejoining by hydrogen bonding
f. salt content less than 100ppm
g. water molecules in the air
h. amount of water vapor in the air
i. amount of water as a percentage of what air can hold at that temperature
j. salt content more than 100ppm
k. containing elements or particulate matter that is beyond the norm
l. process of removing unwanted elements or particulate matter
m. dry region downwind of a mountain
n. all the land area that contributes water to a particular stream or rive
o. non enclosed water impoundments
p. water held in the soil
q. water that soaks into the ground
r. water rising to the atmosphere through evaporation or transpiration and leaving it through condensation and precipitation
s. water movement in the gaseous state promoted by earth's heat or through plants
t. the effect of condensation and cooling temperatures
u. movement of excess noninfiltrated water
v. accumulation of water in an impervious layer of rock or dense clay
w. water subjected to the pull of gravity
x. movement through the soil beyond capillary water
y. the upper surface of accumulated groundwater
z. where water enters an aquifer
aa. layers of porous material through which groundwater moves
bb. where water exits the ground as a significant flow from a relatively small opening

VOCABULARY DEFINITIONS
cc. lost for further human use
dd. results of a collapsed underground cavern drained of it supporting groundwater
ee. gradual settling of land when the water table drops
ff. movement of saltwater into a freshwater aquifer
gg. remains available for further human use
hh. a system of irrigation that applies small steady amount of water to the plant roots
ii. water that has been used for other purposes and reused again
jj. change in the structural characteristics of stream beds
kk. landscaping with desert species that require no additional watering
ll. converting salt into fresh water
mm. movement of water through plants
nn. procedures used to offset the destructive effects of storm water runoff

VOCABULARY EXAMPLES	
a. 0.4%	p. rivulets, streams
b. nonpolluted and < 100ppm salt	q. Ogallala Aquifer
c. water in gaseous state	r. infiltrated water not available to plant roots
d. gas, liquid, solid	s. western Nebraska grasslands
e. water going from gaseous to liquid state	t. irrigation
f. water vapor + temperature	u. Fig. 11.16
g. oceans, seas	v. San Joaquin Valley in California
h. acid rain	w. Fig. 11-17
i. desalinization	x. gray water
j. Death Valley	y. Fig. 11-19
k. Missouri River Basin	z. washing machine, shower water
l. lakes, rivers	aa. Fig. 11-25
m. water available to plant roots	bb. cacti, mesquite bush
n. Fig. 11-3	cc. Fig. 11-20
o. rain, snow	dd. Fig. 11-26

1-15. Illustrate the hydrological cycle including all filtration loops and relevant terms. For example, give yourself one point for each correct representation of: **surface runoff loop, evaporation-transpiration loop, groundwater loop, water vapor, condensation, cloud formation, transpiration, precipitation, evaporation, infiltration, percolation, water table, groundwater, springs, surface runoff.**

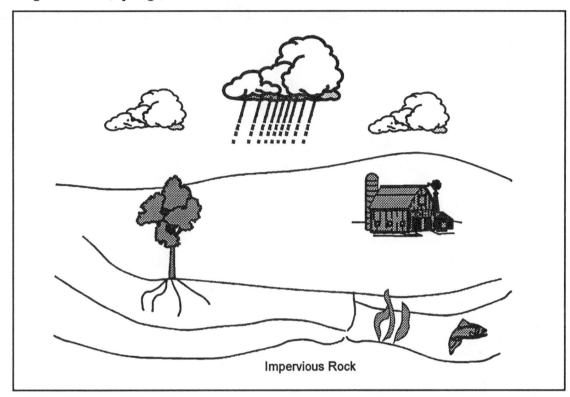

Indicate whether the depletion of groundwater will [+] or will not [-] lead to the following events.

16. [–] increased groundwater use in agriculture
17. [+] subsidence
18. [+] saltwater intrusion
19. [–] increased municipal consumption
20. [+] decreased flow through seeps and springs
21. [+] adverse effects on stream and river ecosystems

As land is developed and natural soil surfaces are replaced by various hard surfaces. As a result of this change indicate whether the following will increase [+], decrease [-], or remain the same [O].

22. [–] amount of groundwater recharge
23. [–] level of water table
24. [–] amount of spring flow
25. [+] stream flow during rains

26.   [   ] frequency of flooding
27.   [   ] pollution of stream channel

Explain how human activities have adversely effected the following filtration loops in the hydrological cycle.

28.   Surface runoff loop: _____

29.   Evapotranspiration loop: _____

30.   Groundwater loop: _____

CHAPTER 12

SEDIMENTS, NUTRIENTS, AND EUTROPHICATION

We all use the word **pollution** and feel we know what it means. Indeed pollution constitutes so many different things that it defies simple definitions. For example, a simple definition such as *"the unfavorable alteration of our surroundings, wholly or largely as a by-product of man's action..."* is so broad and vague that it is virtually meaningless. The only way we can begin to understand the vastness and complexity of the pollution issue is to study the various kinds of things that constitute pollution. In this study, we should become increasingly aware that what we call pollution may be literally hundreds of different kinds of things from as many different sources.

You may want to begin thinking of pollutants in terms of *quantity* and *type*. Quantity is easier to understand than type. Often we expect natural ecosystems to "do something" with thousands of tons of ash, sediment, industrial wastes, and our garbage. Added to these demands, we also ask nature to decompose products that are unrecognizable (e.g., PVC pipe, toxic wastes and other nonbiodegradable materials) to decomposers. In effect, we are asking decomposers to "do their thing" without blueprints or road maps.

Likewise, stopping pollution is not simple. We must come to recognize that wastes are unavoidable by-products of civilization. There is no way that a civilization (or any other population of organisms) can exist without producing wastes. We cannot pretend that we can or will simply stop producing wastes. Instead, the solutions lie in learning to manage and recycle wastes so that they do not cause an "unfavorable alteration of our surroundings." In this chapter and the four that follow it, we will become familiar with major categories of wastes that arise from various sources and we will find that each kind of waste from each source requires its own control/management strategy.

STUDY QUESTIONS

1. The human-caused addition of any material or energy in amounts that cause undesired environmental alterations is called _____.

2. List four undesirable environmental alterations caused by pollution.

   a. _____   b. _____

   c. _____   d. _____

3. List the four actions required to reduce pollution problems.

   a. _____

   b. _____

   c. _____

d. _____

4.   Match the sources of pollutants on the right with categories of pollutants on the left (see Figure 12-1).

Categories   Sources

[   ] toxic chemicals   a. soil erosion and dredging
[   ] nutrient oversupply   b. refuse burning
[   ] sediments   c. coal burning power plants
[   ] particulates   d. cars, trucks, buses, airplanes
[   ] acid forming compounds   e. leaching from lawns and agricultural fields
[   ] fuel combustion products   f. industrial discharges and disposal sites
[   ] pesticides/herbicides   g. fertilizer runoff from several sources

5.   Pollution problems have increased over the years because of increasing _____ and per capita _____ of materials and energy.

The Process of Eutrophication

6.   Loss of the sea grasses in the Chesapeake bay was due to (a. toxic chemicals, b. herbicides, c. turbid water).

7.   Using the terms MORE and LESS, compare the following characteristics of the bay ecology BEFORE and AFTER the massive die off of sea grasses.

Ecological Characteristics	Before	After
a. depth of light penetration	_____	_____
b. amount of sediment	_____	_____
c. amount of phytoplankton	_____	_____
d. dissolved oxygen	_____	_____
e. bacteria	_____	_____
f. photosynthesis	_____	_____
g. fish populations	_____	_____

Different Kinds of Aquatic Plants

8.   Two kinds of aquatic plants are _____ plants which are rooted to the bottom and _____ which mostly consist of single cells or small groups of cells which grow at or near the water surface.

9.   The depth to which light penetrates in sufficient intensity to support photosynthesis is known as the _____ zone.

10. Depending on the turbidity or cloudiness of the water, the euphotic zone may vary from nearly _____ feet in clear water down to only a few inches in very turbid water.

11. Phytoplankton can maintain itself near the surface where there is plenty of light, but it depends on (Surface, Bottom) nutrients.

12.   Benthic plants get their nutrients from the (Surface, Bottom).

13.   The balance between benthic plants and phytoplankton depends on the level of
        _____ in the water?

14.   Do (Benthic Plants or Phytoplankton) depend on light penetration to the bottom to enable
        photosynthesis?

Upsetting the Balance by Nutrient Enrichment

15.   Indicate whether the following properties increase [+] or decrease [-] given the listed
        conditions.

        [+] nutrient ➡ [   ] phytoplankton ➡ [   ] turbidity ➡ [   ] SAV's

The Oligotrophic Condition

16.   The term oligotrophic means nutrient (Rich, Poor).

17.   The two filters of nitrogen and phosphate for natural waterways are:

        a. _____   b. _____

18.   The major source of oxygen in aquatic ecosystems comes from the

        a. atmosphere   b. benthic plants   c. phytoplankton

19.   Indicate whether the following conditions are LOW [-] or HIGH [+] in oligotrophic bodies o
        water. (You will need to read further into the chapter to answer all items.)

        a. [   ] dissolved oxygen   b. [   ] bacterial decomposition

        c. [   ] light penetration   d. [   ] benthic plants

        e. [   ] phytoplankton   f. [   ] water temperature

        g. [   ] water turbidity   h. [   ] nutrient concentrations

        i. [   ] species diversity   j. [   ] aesthetic qualities

        k. [   ] recreational qualities   l. [   ] **BOD biological oxygen demand**

        m. [   ] detritus decomposition   n. [   ] sediments

Eutrophication

20.   The term **eutrophic** means nutrient (Rich, Poor).

21. Phytoplankton do not contribute significantly to dissolved oxygen because

    a. it does not carry on photosynthesis. (True or False)
    b. it does not produce oxygen in photosynthesis. (True or False)
    c. the oxygen it produces escapes from the surface into the atmosphere. (True or False)

22. Does the amount of detritus (Increase or Decrease) in eutrophic water?

23. The increase in detritus comes from the die back of (Benthic Plants, Phytoplankton).

24. Attack of detritus by _____ uses so much _____ that bottom dwelling fish and shellfish suffocate.

25. Decomposers (bacteria) can keep the dissolved oxygen at or near zero for as long as there is detritus to feed on because in the absence of oxygen they can carry on _____ respiration.

26. Indicate whether the following conditions are LOW [-] or HIGH [+] in eutrophic bodies of water. (You will need to read further into the chapter to answer all items.)

    a. [   ] dissolved oxygen   b. [   ] bacterial decomposition

    c. [   ] light penetration   d. [   ] benthic plants

    e. [   ] phytoplankton   f. [   ] water temperature

    g. [   ] water turbidity   h. [   ] nutrient concentrations

    i. [   ] species diversity   j. [   ] aesthetic qualities

    k. [   ] recreational qualities   l. [   ] **BOD biological oxygen demand**

    m. [   ] detritus decomposition   n. [   ] sediments

27. A eutrophic body of water is not technically "dead" because it still supports the abundant growth of _____ and _____.

Eutrophication of Shallow Lakes and Ponds

28. In shallow lakes and ponds, the (SAV, phytoplankton) are the indicators of eutrophication.

Natural Versus Cultural Eutrophication

29. Is it (True or False) that eutrophication of natural bodies of water is part of the natural aging process?

30. List three nutrient sources the cause cultural eutrophication.

    a. _____   b. _____   c. _____

Potential Recovery

31. The key to controlling eutrophication is to reduce _____ inputs.

Additional Factors Contributing to Eutrophication
Impact of Sediments

32. The source of all sediments is _____ erosion.

33. List seven major points of soil erosion and sediments.

    a. _____   b. _____   c. _____

    d. _____   e. _____   f. _____

    g. _____

34. Is it (True or False) that streams, rivers, lakes, and estuaries support complex ecosystems based on many kinds of plant and animal organisms living on or attached to the bottom?

35. Indicate whether the following types of damage to aquatic ecosystems is caused by [a] clay or [b] the bedload of sand and silt.

    [   ] Reduction of light penetration and rate of photosynthesis.
    [   ] Smothers fish by clogging their gills.
    [   ] Scours bottom-dwelling organisms clinging to rocks.
    [   ] Fills in hiding and resting places for fish and crayfish.
    [   ] Causes the stream to become more shallow.

36. _____ is considered the foremost pollution problem of streams and rivers.

37. The impacts of sediments cost the United States $_____ billion each year.

Loss of Wetlands

38. Indicate whether the following are [+] or are not [-] functions of wetlands.

    [   ] Water purification.
    [   ] Groundwater recharge.
    [   ] Filtering nutrients and sediments.
    [   ] Provision of habitat and food for waterfowl and other wildlife.
    [   ] Moderation of wave action.

39. Humans generally have a (Positive, Negative) attitude concerning wetlands.

40. Indicate whether human development projects adjacent to or on wetlands increase [+] or decrease [-] the following wetland functions.

    [   ] Water purification.
    [   ] Groundwater recharge.

[   ] Filtering nutrients and sediments.
[   ] Provision of habitat and food for waterfowl and other wildlife.
[   ] Moderation of wave action.

41.   More than _____ percent of the original wetland acreage in the U.S. has been destroyed.

Global Extent

42.   List two indicators that eutrophication is a global problem.

   a. _____   b. _____

Combating the Symptoms of Eutrophication

43.   List the two general approaches to combating eutrophication.

   a. _____

   b. _____

44.   List four methods that have been used in attacking the symptoms of eutrophication and state the major reason that each is less that successful or practical.

              Method   Shortcoming

   a. _____   _____

   b. _____   _____

   c. _____   _____

   d. _____   _____

Long-term Strategies for Correction

45.   Is it (True or False) that the real control of eutrophication requires decreasing nutrient and sediment inputs?

46.   List three major sources of nutrients that end up in streams and rivers (see Table 12-1).

   a. _____

   b. _____

   c. _____

47.   List seven long-term strategies for controlling eutrophication.

   a. _____

b. _____

c. _____

d. _____

e. _____

f. _____

g. _____

VOCABULARY DRILL

Directions: Match each term with the list of definitions and examples given in the tables below. Term definitions and examples might be used more than once.

Term	Definition	Example (s)	Term	Definition	Example (s)
benthic plants	g	a, p	phytoplankton	f	j, n
BOD	l	b	point-source	r	x
cultural eutrophication	m	c	pollutant	b	l, n
emergent vegetation	i	a, p	pollution	a	m
euphotic zone	j	d	red tides	q	n
eutrophication	d	e	sediment trap	t	o
non-point source	s	f	SAV	h	a, p
nonbiodegradable	c	g	tidal wetlands	o	q, s
nontidal wetlands	p	h	turbid	e	r
oligotrophic	x	i	wetlands	n	h, q, s

VOCABULARY DEFINITIONS
a. human-caused addition of any material or energy in amounts that cause undesired alterations
b. material that causes pollution
c. resist attack and breakdown by detritus feeders and decomposers
d. nutrients rich
e. cloudy or murky
f. free-floating aquatic plants
g. deep rooted aquatic plants
h. aquatic plants that grow entirely under water
i. aquatic plants that grow partially under and above water
j. depth of adequate light for photosynthesis
k. nutrient poor
l. measure of pollution level by ppm oxygen required for decomposition
m. accelerated eutrophication caused by humans
n. land areas naturally covered by water for specified periods of time
o. coastal wetlands
p. inland wetlands
q. indicators of oceanic eutrophication
r. a fixed-point pollutant source
s. a diffuse pollutant source
t. a means of controlling erosion from construction sites

VOCABULARY EXAMPLES	
a. cattails, water sedges	k. sewage treatment plant
b. ppm dissolved O2	l. nitrate and phosphate
c. Fig. 12-5	m. eutrophication
d. 1 inch to 600 feet	n. marine phytoplankton
e. Chesapeake Bay	o. Fig. 12-16
f. urban runoff	p. Fig. 12-2a
g. synthetic organic compounds	q. Fig. 12-8
h.. Fig. 12-9a	r. most rivers in the U.S.
i. wetlands	s. Florida Everglades
j. filamentous or single-cell algae	

SELF-TEST

Indicate whether the following conditions are characteristic of [O] oligotrophic or [E] eutrophic bodies of water.

1. [   ] high dissolved oxygen

2. [   ] low light penetration

3. [   ] abundant phytoplankton

4. [   ] low water turbidity

5. [   ] high species diversity

6. [   ] high recreational qualities

7. [   ] high detritus decomposition

8. [   ] high bacterial decomposition

9. [   ] abundant benthic plants

10. [   ] cool water temperature

11. [   ] high nutrient concentrations

12. [   ] poor aesthetic qualities

13. [   ] low BOD

14. [   ] low sediments

List four major sources of sediments and nutrients that end up in streams and rivers.

Sediments

15. _____

17. _____

19. _____

21. _____

Nutrients

16. _____

18. _____

20. _____

22. _____

133

Explain how each of the following events or materials contribute to **eutrophication**.

23.   Sedimentation: _____

_____

24.   Nutrients: _____

_____

25.   Decomposition: _____

_____

26.   Which of the following approaches to combating eutrophication is getting at the root cause?

a. chemical treatments   b. decreasing nutrient input   c. aeration   d. harvesting algae

27.   The foremost pollution problem of streams and rivers is

a. toxic wastes   b. salinization   c. sedimentation   d. eutrophication

28.   Nutrients flow into aquatic ecosystems from

a. fertilizers from croplands.
b. animal wastes from feedlots.
c. detergents containing phosphate.
d. all of the above

29.   Lands that are naturally covered by shallow water at certain times and more or less drained at others are called

a. grasslands   b. wetlands   c. rain forests   d. estuaries

Indicate whether the following functions of wetlands will increase [+] or decrease [-] after the establishment of human "development" projects on or adjacent to wetlands.

30.   [–] water purification.

31.   [–] provision of habitat and food for waterfowl and other wildlife.

32.   [–] groundwater recharge

CHAPTER 13

SEWAGE POLLUTION AND REDISCOVERING THE NUTRIENT CYCLE

Human sewage is one pollutant that over the course of human history has caused more deaths and cases of illness than all other pollutants combined. This pollutant is now largely under control in developed countries, but it still causes much illness and many deaths each year in developing countries.

In natural ecosystems, the breakdown of excrements and the reabsorption of the elements they contain is a crucial aspect of all the nutrient cycles. However, it is also possible to "recycle" various parasites and disease-causing organisms while we are recycling the nutrients in excrements. This can lead to great epidemics of disease, suffering and death. Recall that in natural ecosystems such epidemics are a major factor in controlling populations, but this kind of crude control we wish to avoid in human society.

The work of Louis Pasteur and others in the mid 1800s led us to recognize the disease hazards associated with excrements. Consequently, sewage collection and disposal systems were developed. These were the flush-away-with-water systems, the same concept that we still use today. These systems greatly reduce the disease hazard by removing the excrements from the immediate vicinity of human habitation. Thus, they reduce the potential of contaminating water and food supplies with the disease-causing organisms that may be present. Unfortunately, such systems also introduce large loads of nutrients into bodies of water. This transfer is a one-way street because as water evaporates the nutrients remain behind. Even the most widely used sewage treatment processes do not remove the nutrients from the water. The consequence is that receiving bodies of water are overloaded with nutrient inputs, which leads to eutrophication and the loss of nutrients from the land-based cycle. It is now possible to have disease control, avoidance of pollution due to nutrient overloading, and recycling of nutrients.

STUDY QUESTIONS

Health Hazard of Untreated Sewage

1. The major problem resulting from untreated sewage wastes is the spread of infectious
   _____.

2. Disease causing bacteria, viruses, and other parasitic organisms that infect humans and other animals are called _____.

3. List four diseases caused by the pathogens present in sewage (see Table 13-1).

   a. _____   b. _____

   c. _____   d. _____

4. Transmission of pathogens in infected sewage waste may occur in what three ways?

   a. _____   b. _____   c. _____

5. The level of infection of human populations (Is, Is Not) related to population density.

6. Public health measures which prevent a disease cycle involve

    a. _____

    b. _____

    c. _____

Development of Collection and Treatment Systems

7. Prior to the late-1800s, the general means of disposing of human excrements was the
    _____ privy which often was located near a _____.

8. What person showed the relation between sewage-borne bacteria and infectious diseases?
    _____

9. The earliest forms of sewage disposal transferred it directly into _____ drains.

10. _____ societies initiated the system of flushing sewage into natural
    waterways.

11. Around 1900, sewage treatment included _____ the wastewater and
    separating _____ and _____ water.

12. Is it (True or False) that there are no longer any cases of raw sewage overflowing with
    stormwater into waterways?

Sewage Management and Treatment
The Pollutants in Raw Sewage

13. The total of all the drainage from sinks, bath tubs, laundry, and toilets constitutes a mixture
    called raw _____.

14. Raw sewage is actually _____ percent water and _____ percent pollutants.

15. List the four major categories of pollutants in raw sewage and give an example of each.

            Category   Example

    a. _____   _____

    b. _____   _____

    c. _____   _____

    d. _____   _____

Removing the Pollutants From Sewage

16.   The two major sewage treatment problems in the United States are those of dealing with:

        a. _____ and   b. _____

17.   Match the below list of wastewater treatment technologies or products with [a] preliminary treatment, [b] primary treatment, [c] secondary treatment, [d] biological nutrient removal.

        [   ] primary clarifiers
        [   ] bar screen
        [   ] grit settling tank
        [   ] raw sludge
        [   ] biological treatment
        [   ] activated sludge system
        [   ] activated sludge
        [   ] rotating screens
        [   ] course sand and gravel
        [   ] removal of dissolved nutrients
        [   ] uses decomposers and detritus feeders
        [   ] settling of raw sludge
        [   ] aeration

Preliminary Treatment

18.   First, large pieces of debris are removed by passing the water through a _____ screen.

19.   Coarse sand and other dirt particles are removed by passing the water through a _____ settling tank where the velocity of the water is (Increased, Decreased) and the particles are allowed to _____ out.

20.   The large debris from the bar screen are taken out and _____.

21.   Grit from the grit settling tank is put in _____.

Primary Treatment

22.   In primary treatment

        a. the water flows into very large tanks called _____ **clarifiers.**
        b. the water flows through (Quickly, Slowly).
        c. heavier particles of organic matter (Disperse, Settle Out).

23.   What percentage of the organic material is removed by this settling process? _____

24.   The material that settles out is called **raw** _____.

Secondary Treatment

25. Secondary treatment is also called _____ treatment because decomposers and detritus feeders are used.

26. In the activated sludge system, a mixture of organisms is added to the water to be treated and it is passed through a tank where it is _____.

27. In the aeration tank, the organisms feed on _____ matter.

28. Following the aeration tank, the water is passed on to a secondary _____ tank where the organisms are returned to the _____ tank.

29. Through the activated sludge systems, _____ percent of the organic material is removed.

Biological Nutrient Removal (BNR)

30. BNR utilizes aspects of the _____ and _____ cycles.

Final Cleansing and Disinfection

31. Indicate whether the following statements represent positive [+] or negative [-] attributes of using chlorine gas as the means of killing pathogenic bacteria in wastewater.

    [   ] Cost
    [   ] Toxicity during transportation and to fish.
    [   ] Chlorinated hydrocarbons.

32. Two alternative technologies for killing pathogenic bacteria in wastewater are

    a. _____ and b. _____

Sludge Treatment Options

33. Raw sludge is about _____ percent water and _____ percent organic matter.

34. Four sludge treatment methods are

    a. _____ b. _____

    c. _____ d. _____

35. **Anaerobic digestion** of raw sludge in the absence of _____ produces the products of _____ gas and _____ **sludge.**

36. Composting basically produces _____.

37. Pasteurization and drying produce sludge _____.

38.   Lime stabilization uses a mixture of sludge and _____.

39.   Indicate whether the following statements indicate progress [+] or the lack of [-] progress in waste water treatment.

[   ] Waste water treatment in less developed countries.
[   ] Sewage being discharged with stormwater into natural waterways.
[   ] Increasing population overburdening existing systems.
[   ] Clean Water Act of 1972.
[   ] Communities discharging raw sewage into bays, lakes, and coastal waters.
[   ] Dumping sludge into landfills or the sea.

Alternative Treatment Systems

40.   List five alternative methods of extracting wastewater nutrients.

a. _____   b. _____

c. _____   d. _____

e. _____

Individual Septic Systems

41.   The two basic component to individual septic systems are a _____ tank and _____ field.

42.   Prerequisites for septic systems are enough _____ and subsoil that allows _____ of water.

Impediments to Recycling Sewage Products

43.   List two major impediments to progress in waste water treatment.

a. _____

b. _____

44.   Is it (True or False) that toxic chemical wastes can be removed through advanced treatment?

Directions: Match each term with the list of definitions and examples given in the tables below. Term definitions and examples might be used more than once.

Term	Definition	Example (s)	Term	Definition	Example (s)
activated sludge			primary clarifiers		
bar screen			primary treatment		
biogas			raw sludge		
biological treatment			raw sewage		
BNR			sanitary sewers		
chlorinated hydrocarbons			secondary treatment		
co-composting			settling tank		
composting toilet			sludge cake		
pasteurized			sludge digester		
pathogens			storm drains		
preliminary treatment			treated sludge		

VOCABULARY DEFINITIONS

a. sewage with no treatment whatsoever

b. disease-causing organisms

c. system to collect runoff from precipitation

d. system to collect all wastewater

e. first stage in the removal of pollutants from sewage

f. devise to remove large objects from wastewater

g. devise to remove sand and gravel from wastewater

h. first stage in the removal of organic matter from wastewater

i. devise that allows organic matter to settle out of wastewater

j. material removed during primary treatment

k. stage in wastewater treatment that uses natural decomposers and detritus feeders

l. a mixture of sludge and detritus-feeding organisms

VOCABULARY DEFINITIONS
m. stage in wastewater treatment that removes dissolved nutrients
n. result of spontaneous reactions between chlorine and organic compounds
o. devise that allows bacteria to feed on detritus in the absence of oxygen
p. a major byproduct from the sludge digester
q. a dehydrated form of treated sludge
r. a mixture of raw sludge and shredded waste paper
s. procedure for killing pathogens
t. a waterless toilet

VOCABULARY EXAMPLES	
a. Fig. 13-8	k. particulate organic matter and fatty material
b. methane	l. fluid that enters the sewage treatment plant
c. Fig. 13-9	m. plumbing from sinks, tubs, and toilets
d. cancer-causing compounds	n. Fig. 13-7
e. Fig. 13-14	o. Fig. 13-5
f. Cilvus Multrum	p. Fig. 13-13
g. Milorganite	q. Fig. 13-11
h. bacteria and viruses	r. street curb drains
i. Fig. 13-3	s. nutrient-rich humus like material
j. Fig. 13-6	

SELF-TEST

The two major problems resulting from untreated sewage wastes going into waterways are

1. _____   2. _____

√ 3.   The primary categories of pollutants in raw sewage are

   a. pathogens, detritus, and dissolved nutrients.
   b. phosphates, nitrates, and sulfates.
   c. heavy metals, synthetic organisms, and pesticides.
   d. industrial, household, and municipal wastes.

4.   What proportion of raw sewage is actually pollutant? _____ %

        Match the below list of wastewater treatment technologies or products with [A] preliminary treatment, [B] primary treatment, [C] secondary treatment, or [D] BNR.

5. [   ] irrigation   6. [   ] bar screen   7. [   ] aeration

8. [   ] raw sludge   9. [   ] denitrifying bacteria   10. [   ] activated sludge

        Match each of the following facilities [a to d] with the function (questions 11 to 14) it performs.

        a. activated sludge system, b. bar screen, c. primary clarifier, d. none of these

11.   [   ] Screens out large pieces of debris.

12.   [   ] Enables microorganisms to digest organic matter present.

13.   [   ] Allows 50 to 70 percent of the organic matter to settle out.

14.   [   ] Removes most of the dissolved nutrients.

15.   Which of the following sludge treatments would not produce methane gas?

        a. anaerobic digestion   b. composting   c. pasteurization   d. lime stabilization

16.   Most sewage treatment plants do not have

        a. pretreatment, b. primary treatment, c. secondary treatment, d. BNR

17.   Explain why conventional methods of wastewater treatment only partially solve the problem of water pollution.

        _____

18.   The treated water from most sewage treatment plants contains

        a. nutrients   b. bacteria   c. sediments   d. colloidal materials

        List two major impediments to progress in waste water treatment.

19.   _____

20.   _____

142

CHAPTER 14

POLLUTION FROM HAZARDOUS CHEMICALS

Before you begin studying Chapter 14, imagine someone approaching you and making the following proposal. This person says that he/she represents a major industry in the community that has provided hundreds of jobs to the residents. These jobs and the products produced by this industry have given economic security to the community. However, the industry is having some difficulty disposing of the toxic wastes resulting from product production. The industry is going to bury or discard the toxic wastes in an undetermined location with as little cost to the companies stockholders as possible. How would you react to this proposal? What questions do you need to have firm answers to before you would support the proposal? Even though the above problem is occurring all over the United States, the way it has been presented to you is really a hypothetical situation. It is hypothetical because you would not be given the opportunity to react to the proposal but you would be given the opportunity to suffer the consequences.

Groundwater has traditionally been a major source of high quality water, suitable for drinking without further treatment or purification. Water supplies for about half the population of the United States are drawn directly from groundwater through personal or municipal wells. Assuming we keep withdrawal in balance with recharge, groundwater could last forever, a renewable resource of immeasurable value. But now groundwater supplies are, with increasing frequency, being found to be polluted with toxic chemicals which may have insidious health effects. These chemicals are not natural products. They are products and by-products of human activities. In short, we are poisoning our own well.

What are these chemicals? Where do they come from? What can we do to avoid contaminating ourselves with them? These are the questions which we will be addressing in this chapter.

STUDY QUESTIONS

1. What types of chemical disasters occurred at the following locations.

    a. Cuyahoga River: _____

    b. Minimata, Japan: _____

    c. Bopal, India: _____

    d. Niagra, NY: _____

The Nature of Chemical Risks: HAZEMATs

2. Match the examples of hazardous materials on the left with the EPA category on the right.

    [ ] gasoline   a. toxicity
    [ ] acids   b. reactivity

[ ] explosives   c. corrosivity

[ ] pesticides   d. ignitability

3.   List five environmental costs in the **product life cycle** of your shampoo.

a. _____

b. _____

c. _____

d. _____

e. _____

4.   Is it (True or False) the predominant source of chemicals entering the environment are from households?

Toxic Chemicals Presenting a Long-Term Threat

5.   The two major classes of chemicals that are particularly significant environmental pollutants are _____ and _____.

Heavy Metals

6.   Examples of heavy metals are:

a. _____   b. _____   c. _____

d. _____   e. _____   f. _____

g. _____   h. _____

7.   How are heavy metals used in industry?

_____

8.   Heavy metals are extremely toxic because they combine with and inhibit the functioning of _____ causing _____ and _____ effects.

Nonbiodegradable Synthetic Organics

9.   Synthetic organic chemicals are the basis for the production of all _____, synthetic _____, synthetic _____, _____ coatings, _____, _____, and _____ preservatives.

10.   Synthetic organic chemicals are (Biodegradable, Nonbiodegradable).

11. List three health effects from synthetic organics:

    a. _____ b. _____ c. _____

12. A particularly dangerous group of synthetic organic compounds are the _____ **hydrocarbons**.

13. The main feature of halogenated hydrocarbons is that one or more hydrogen atoms have been substituted by an atom of _____, _____, _____ or _____.

14. The most common halogenated hydrocarbons are those containing chlorine. Such compounds are referred to as _____ **chlorides**.

15. Examples of chlorinated hydrocarbons include:

    a. _____, b. _____, c. _____

    d. _____, and e. _____.

Bioaccumulation and Biomagnification

16. Both heavy metals and chlorinated hydrocarbons are particularly insidious because they tend to _____.

17. Bioaccumulation of these compounds occurs because they are

    a. biodegradable. (True or False)
    b. nonbiodegradable. (True or False)
    c. readily absorbed by the body. (True or False)
    d. readily excreted from the body. (True or False)

18. Effectively, the body _____ synthetic organics.

19. Bioaccumulation occurs in the _____ organism whereas biomagnification occurs in the _____ _____.

20. Biomagnification demonstrates how each higher trophic level receives and accumulates a (higher, lower, the same) dose than the one before. (see Figure 14-3)

21. Through a food chain the concentration of a toxic substance in organisms may be increased by as much as a _____ times.

Interaction Between Pollutants and Pathogens

22. Match [a] DDT, [b] interaction between pollutants and pathogens, and [c] mercury with the bioaccumulation episode each caused.

    [   ] Minamata Disease   [   ] Diebacks of predatory birds   [   ] dolphin deaths

A History of Mismanagement
Indiscriminate Disposal

23. Prior to the passing of environmental laws in the 60s and 70s, it was common practice to dispose of chemical wastes into the _____ or natural _____.

24. The public outcry against pollution in the 60s led the U.S. Congress to pass two major pieces of legislation called:

   a. _____ Date _____

   b. _____ Date _____

25. In the years since these laws were passed direct discharges of wastes into natural waterways and the air have (Decreased, Increased).

26. Decreases of waste discharges has led to (more, less) industrial waste material requiring some means of disposal.

Methods of Land Disposal

27. Indicate whether the following statements describe the disposal technique of [a] deep well injection, [b] surface impoundments, or [c] landfills.

   [   ] Concentrated wastes put in drums and buried.
   [   ] A reverse well.
   [   ] An open pond without an outlet.
   [   ] Volatile wastes may enter the atmosphere.
   [   ] Waste put into deep rock strata.
   [   ] Involves the removal of leachate.

28. Indicate whether the following problems of improper disposal pertain to [a] deep well injection, [b] surface impoundments, [c] landfills, or [d] all three.

   [   ] Absence of liners or leachate collection systems.
   [   ] Deposition of wastes above or into aquifers.
   [   ] Immersion of wastes into ground water.
   [   ] Lack of reliable monitoring systems to detect leakage.

29. Is it (True or False) that once wastes have seeped into groundwater, there is no practical way to remove them?

30. Small amounts of toxic chemicals in groundwater are extremely hazardous because of their known ability to _____ and cause serious _____ _____.

31. Is it (True or False) that when land disposal facilities are properly sited, constructed, and maintained, they guarantee that wastes will never seep or leach into groundwater?

32. Two problems inherent in land disposal is that wastes _____ at disposal facilities and that they _____ there.

Scope of the Problem

33. The use of designated disposal sites that have no precautions against ground pollution are examples of _____ land disposal. Give an example. _____

34. There are an estimated _____ million tons of hazardous wastes generated each year in the United States.

35. List the three major aspects to the toxic waste problem.

    a. _____

    b. _____

    c. _____

Cleaning Up the Mess
Ensuring Safe Drinking Water

36. A federal law that attempts to assure safe drinking water supplies is the
    _____ Date _____

37. Indicate whether the following are [+] or are not [-] provisions of the Safe Drinking Water Act of 1974.

    [   ] Setting standards regarding allowable levels of pollution.
    [   ] Monitoring municipal water supplies.
    [   ] Proper location and construction of injection wells.
    [   ] Systematic monitoring of groundwater and private wells.
    [   ] Determination of "safe" and "unsafe" levels for all chemicals.
    [   ] Provision of funds to compensate for illness resulting from contaminated wells.

38. Identify two emerging issues in the public debate over the Safe Drinking Water Act of 1974.

    a. _____

    b. _____

Superfund for Toxic Sites

39. The **Comprehensive Environmental Response, Compensation, and Liability Act of 1980** is commonly known as the _____.

40. How is this act funded? _____

41. The Superfund for 1980 to 1985 was $_____ billion.

42.   The Superfund for 1986 to 1991 was $_____ billion.

43.   The Superfund for 1991 to 1996 is $_____ billion.

44.   What types of actions are funded by the Superfund?

       a. _____

       b. _____

       c. _____

       d. _____

45.   The administrative unit for the Superfund is the _____.

46.   Indicate whether the following have been positive [+] or negative [-] attributes of the EPA's "administration" of the Superfund.

       [   ] Placement of NPL sites on the superfund list.
       [   ] Scope of action across all NPL sites in the United States.
       [   ] Development of technologies to clean-up NPL sites.
       [   ] Cost of cleaning-up NPL sites.
       [   ] Cost of administering the Superfund

47.   Is it (True or False) that once groundwater is contaminated, it is effectively lost forever?

Management of New Wastes

48.   Indicate whether the following have been positive [+] or negative [-] attributes of the administration of the Clean Water Act.

       [   ] Interim permits
       [   ] Monitoring system
       [   ] Paying fines
       [   ] Exemptions

49.   Is it (True or False) that the Clean Water Act has stopped nearly all dumping of toxic wastes into natural waterways?

Regulating Land Disposal

50.   The RCRA legislation required

       a. permitting of disposal sites. (True or False)
       b. pretreatment of wastes destined for landfills. (True or False)
       c. cradle to grave tracking of all hazardous wastes. (True or False)

51. Indicate whether the following are known [+] or anticipated [?] results of the RCRA legislation.

[   ] The closure of many waste disposal facilities.
[   ] Disposal of wastes in unregulated land fills.
[   ] Legally permitted discharges.
[   ] New methods of midnight dumping.
[   ] Tougher laws in some states.
[   ] Shipment of hazardous wastes to other countries.
[   ] Wholesale movement of some industries to other countries.

Reduction of Accidents and Accidental Exposures

52. Match the four potential sources of accidental exposures to hazardous wastes with the safeguards or legislation that have been developed to prevent accidental exposures.

Sources of Exposure   Safeguards

[   ] leaking underground storage tanks   a. DOT regs
[   ] transportation accidents   b. SARA
[   ] workplace accidents   c. TSCA
[   ] new chemicals   d. UTS

Looking Toward the Future: Pollution Avoidance

53. The EPA estimates of annual costs of pollution control and clean-up programs is $_____ billion.

Too Many or Too Few Regulations?

54. There are _____ billion pounds of toxic chemicals discharged into the environment annually.

55. Two groups that are usually exempted from toxic waste regulation compliance are _____ and _____ .

56. Why are these groups exempted? _____

Pollution Avoidance for a Sustainable Society

57. List four general approaches in pollution prevention programs.

a. _____

b. _____

c. _____

d. _____

VOCABULARY DRILL

Directions: Match each term with the list of definitions and examples given in the tables below. Term definitions and examples might be used more than once.

Term	Definition	Example (s)	Term	Definition	Example (s)
bioaccumulation			ground water remediation		
biomagnification			NPL		
bioremediation			organic chlorides		
halogenated hydrocarbons			pollution avoidance		
HAZMAT			pollution control		
midnight dumping			product life cycle		

VOCABULARY DEFINITIONS

a. a chemical that presents a hazard or a risk

b. cradle to grave product history

c. organic compounds in which one or more hydrogen atoms have been replaced by halogens

d. chlorinated hydrocarbons

e. when harmless amounts of toxic waste received over long periods of time reach toxic levels

f. multiplying effect of toxic wastes through the food chain

g. illegal deposition of toxic wastes in any location available under the cover of darkness

h. a technique to clean contaminated groundwater

i. sites representing the most immediate and severe toxic waste threats

j. utilization of oxygen and organisms to clean-up toxic waste sites

k. adding filters to prevent environmental pollution

l. changing the production process so that harmful pollutants are not produced

VOCABULARY EXAMPLES	
a. electric cars	g. CCl4
b. biodegradable toxic organic compounds	h. Fig. 14-1
c. catalytic converter	i. military bases
d. Fig. 14-8	j. Minamata disease
e. Fig. 14-3	k. Fig. 14-12
f. gasoline, acids, explosives	

SELF-TEST

Give an example of a hazardous chemical that fits each of the categories listed below.

1.   Toxicity: _____

2.   Reactivity: _____

3.   Corrosivity: _____

4.   Ignitability: _____

Match the sources of chemicals entering the environment with an example.

Sources

5.	[ ] Entire products	a. Tanker trucks
6.	[ ] Fractional products	b. UST's
7.	[ ] Waste byproducts	c. Your discarded shampoo bottle
8.	[ ] Household solvents	d. Cleaning fluids
9.	[ ] Unused portions	e. Heavy metals
10.	[ ] Leaks	f. Paint solvents
11.	[ ] Accidental spills	g. Fertilizers

Examples

12.   The two categories of wastes that are particularly toxic are

   a. midnight dumping and inadequate landfills.
   b. heavy metals and synthetic organics.
   c. deicing salt and waste road oil.
   d. sewage sludge and transportation spills.

13.   _____ is the term defining organisms' inability to excrete heavy metals.

14.   _____ is the term defining the multiplying effect of concentrations of heavy metals in the food chain.

Indicate whether the following statements describe the disposal technique of [a] deep well injection, [b] surface impoundments, [c] landfills, or [d] all three

15.   [   ] Concentrated wastes put in drums and buried.
16.   [   ] A reverse well.
17.   [   ] An open pond without an outlet.
18.   [   ] Volatile wastes may enter the atmosphere.
19.   [   ] Waste put into deep rock strata.
20.   [   ] Involves the removal of leachate.
21.   [   ] Absence of liners or leachate collection systems.
22.   [   ] Deposition of wastes above or into aquifers.
23.   [   ] Immersion of wastes into ground water.
24.   [   ] Lack of reliable monitoring systems to detect leakage.

25.   The federal program aimed at identifying and cleaning up existing toxic waste sites is the

        a. Resources Conservation and Recovery Act of 1976
        b. Clean Water Act of 1974
        c. FIFRA
        d. Comprehensive Environmental Response, Compensation, and Liability Act of 1980

Which of the following legislative acts were designed to control disposal of chemical (toxic) wastes. (Check all that apply)

26.   [   ] Clean Air Act of 1970
27.   [   ] Clean Water Act of 1972
28.   [   ] Safe Drinking Water Act of 1974
29.   [   ] Comprehensive Environmental Response, Compensation, and Liability Act of 1980
30.   [   ] Resource Conservation and Recovery Act of 1976
31.   [   ] Toxic Substances Control Act of 1978

Distinguish the difference in pollution avoidance and control.

32.   Avoidance: _____

33.   Control: _____

CHAPTER 15

AIR POLLUTION AND ITS CONTROL

▬▬▬▬▬▬▬▬▬▬▬▬▬▬▬▬▬▬▬▬▬▬▬▬▬▬▬▬▬▬▬▬▬

Natural forest fires! Volcanic eruptions! Dust storms! Such events have always thrown foreign particles and gases (pollutants) into the atmosphere. But, there are also natural purification processes at work. Larger particles settle of their own accord and smaller particles and water soluble gases are continually cleansed from the atmosphere by the water cycle. Water vapor condenses on particles and gases may dissolve in the forming droplet and, thus, both are brought down with precipitation. Some foreign gases may be absorbed and assimilated by vegetation and soil microorganisms and if the amounts are not too great, the organisms do not suffer. Since natural events causing air pollution are generally few and infrequent, the natural biosphere maintained a balance between pollution input and removal in which the balance generally rested on clean air.

But now each year in the United States alone, we humans burn some 5 billion barrels of oil, 1 billion tons of coal, 22 trillion cubic feet of natural gas, mountains of refuse, and then add to this the materials from metal refining and all the fumes from organic chemicals that are produced in their manufacture and use. What happens to the balance? Indeed, that the air is not much worse than it is speaks to the remarkable capacity of natural processes and some to our own efforts in pollution control. However, should we be satisfied and complacent? Forests in many areas are dying from pollution, and in some areas pollution indexes reach unhealthful if not hazardous levels. Clearly, vigilance is in order and we need to do more to curtail our emissions of pollutants into the atmosphere. This chapter will describe what the major air pollutants are, where they come from, and what is being done and remains to be done to control them.

▬▬▬▬▬▬▬▬▬▬▬▬▬▬▬▬▬▬▬▬▬▬▬▬▬▬▬▬▬▬▬▬▬

STUDY QUESTIONS

Background of the Air Pollution Problem

Structure of the Atmosphere

1. Indicate whether the following statements pertain to the [a] **trophosphere**, [b] **tropopause**, or [c] **stratosphere**.

    [   ] contains ozone
    [   ] extends up 10 miles
    [   ] caps the trophospere
    [   ] gets colder with altitude
    [   ] temperature increases with altitude
    [   ] well mixed vertically
    [   ] site and source of our weather
    [   ] substance that enter remain there a long time
    [   ] pollutants can reach the top within a few days

Atmospheric Pollution and Cleansing

2.   List two processes that hold natural pollutants below the toxic level.

      a. _____

      b. _____

3.   Is it (True or False) that organisms do have the capacity to deal with certain amounts or levels of pollutants without suffering ill effects?

4.   Pollution effect depends on _____ and _____ of exposure.

5.   Indicate whether the **threshold level** is high [+] or low [-] for:

      [   ] short exposure time   [   ] long exposure time

6.   Is it (True or False) that there are some compounds that have no threshold level?

7.   Dose is defined as _____ multiplied by time of _____.

8.   List the three factors that determine air pollution levels

      a. _____   b. _____

      c. _____

The Appearance of Smogs

9.   The discovery of _____ began human inputs of air pollutants.

10.   In Hard Times, Charles Dickens referred to _____ **smog**.

11.   Thousands of cars venting unprotected exhaust + sunlight + a mountain topography = _____ **smog**.

12.   Normally temperature (Increases, Decreases) as elevation increases.

13.   The warmer surface air _____ which tends to carry pollutants away.

14.   Warmer surface air cannot rise when a layer of (Warmer, Cooler) air overlies warmer air, a situation that is referred to as a _____ **inversion**.

15.   What is the highest recorded number of human deaths due to an *air pollution disaster*? _____

16.   Air pollution may

      a. adversely affect human health. (True or False)
      b. cause damage to crops and forests. (True or False)

c. increase the rate of corrosion and deterioration of materials. (True or False)

17.   Is it (True or False) that human illnesses and damage to crops and forests from air pollution has become increasingly commonplace?

The Clean Air Act

18.   A major law passed by the U.S. Congress which addresses the problem of air pollution is the _____ of _____ and its amendments of _____ and _____.

19.   The standards of the Clean Air Act of 1970

   a.   set the levels of air pollution control needed to protect the environment and human health. (True or False)
   b.   set the time tables for meeting pollution control levels. (True or False)

20.   Identify the four steps involved in meeting the mandates of the Clean Air Act of 1970.

   a. _____

   b. _____

   c. _____

   d. _____

21.   Is it (True or False) that air quality in most U.S. cities is better now than it was in the early 1970s?

Major Air Pollutants and Their Impact
Major Pollutants

22.   List the eight pollutants or pollutant categories that are the most widespread and serious.

   a. _____ b. _____ c. _____

   d._____ e. _____ f. _____

   g. _____ h. _____

23.   In large measure, all of the above pollutants are direct or indirect products of fossil fuel

   _____

Adverse Effects of Air Pollution on Humans, Plants, and Materials

24.   Air pollution

   a. is the result of one chemical mixed with the normal constituents of air. (True or False)
   b. varies in time and place. (True or False)

c. varies with environmental conditions. (True or False)

25.   Is it (Easy or Difficult) to determine the role of particular pollutants in causing an observed result?

Effects on Human Health

26.   Identify the three categories of air pollutant impacts on humans.

   a. _____   b. _____   c. _____

27.   Serious adverse effects of air pollution on human health are seen

   a. mainly among smokers b. mainly among nonsmokers c. equally among both groups

28.   Is it (True or False) that smoking by itself greatly increases the risk of serious disease?

29.   Learning disabilities in children and high blood pressure in adults are correlated with high levels of _____ in the blood.

30.   The source of this widespread contamination is

   a. people who smoke b. use of leaded gasoline c. high ozone levels d. pesticide use

31.   Which country has the worst air in the world? _____

32.   The EPA mandated the phase-out of leaded gasoline by the end of 1988. Is it (True or False) that you can still buy leaded gasoline in 1995?

Effects on Agriculture and Forests

33.   Air pollution

   a. may destroy vegetation. (True or False)
   b. may seriously retard the growth of crops and forests. (True or False)
   c. may cause forests to become more vulnerable to insect pests and disease. (True or False)

34.   Indicate whether the following effects of air pollutants were [+] or were not [-] demonstrated in open-chamber experiments.

   [   ] Which pollutants cause the damage.
   [   ] The sensitivity level of plants to gaseous air pollutants.
   [   ] The degree to which air pollutants retard plant growth and development.
   [   ] The degree to which air pollutants reduce crop yields and profits.

35.   The most serious pollutant effecting agriculture and natural areas is _____.

36.   Is it (True or False) that crops and wild plants react in the same manner to air pollutants?

37. Is it (True or False) that just a small increase in concentration or duration of exposure may push some plants beyond their ability to cope with an air pollutant?

Effects on Materials and Aesthetics

38. Indicate whether the following conditions could [+] or could not [-] be an effect of air pollution.

    [   ] Grey and dingy walls and windows.
    [   ] Deteriorated paint and fabrics.
    [   ] Corrosion of metals.
    [   ] A bright, clear, blue sky.
    [   ] Increased real estate values.
    [   ] Decreased real estate values.

Pollutant Sources

39. Air pollutants are direct or indirect by-products from burning _____,
    _____ and _____.

40. Oxidation of coal, gasoline, and refuse by burning (Is, Is Not) usually complete.

Primary and Secondary Pollutants

41. Indicate whether the following are primary [P] or secondary [S] pollutants from combustion.

    [   ] sulfuric acid
    [   ] hydrocarbon emissions
    [   ] ozone
    [   ] nitric oxide
    [   ] sulfur dioxide
    [   ] carbon monoxide
    [   ] PANs
    [   ] particulates
    [   ] nitrogen dioxide
    [   ] lead
    [   ] **photochemical oxidants**
    [   ] volatile organic compounds (VOC's)

√ 42. Which pollutant causes the brownish color of photochemical smog? _____

43. Identify the major source(s) of: (see Figure 15-14)

    Particulates: _____   All other air pollutants: _____

Bringing Air Pollution Under Control

Setting Standards

44. The law that mandated setting standards for air pollutants is the _____.

157

45.   The Clean Air Act mandated setting standards for

a. all air pollutants. (True or False)
b. six pollutants are recognized as most widespread and objectionable. (True or False)

46.   List the six pollutants that are most widespread and objectionable.

a. _____   b. _____   c. _____

d. _____   e. _____   f. _____

Control Strategies

47.   Explain the **command-and-control** approach.

_____

_____

48.   Indicate the level, high [H] or low [L], of success of the command-and-control approach for the following air pollutants.

a. [   ] lead   b. [   ] ozone   c. [   ] particulate matter   d. [   ] SO2

49.   Lead emissions have been reduced by _____ percent in the past 25 years.

Reducing Particulates

50.   Identify two devices that some industries have installed to reduce particulates.

a. _____   b. _____

51.   Is it (True or False) that these devices remove all toxic substances?

52.   List four contemporary sources of particulates.

a. _____   b. _____

c. _____   d. _____

53.   How does the Clean Air Act of 1990 address the 83 regions of the U.S. that have failed to attain air particulates standards?

_____

Limiting Pollutants from Motor Vehicles

54.   A new car today emits _____ percent less pollutants than pre-1970 cars.

55. Explain how a **catalytic converter** works.

_____

_____

56. List the three major considerations of the Clean Air Act of 1990 concerning motor vehicle pollutants.

a. _____

b. _____

c. _____

Decreasing Sulfur Dioxide and Acids

57. The major source of sulfur dioxide emissions is _____.

58. Approximately _____ tons of sulfur dioxide/day can be emitted for every 10,000 tons of coal burned/day.

59. What is the goal of the Clean Air Act of 1990 concerning sulfur dioxide emission from coal-burning plants by the year 2010?

_____

Managing Ozone

60. Identify a **point** and **area** source of VOC emissions.

Point: _____   Area: _____

61. List two VOC control strategies in the Clean Air Act of 1990.

a. _____

b. _____

Controlling Toxic Chemicals in the Air

62. The estimated total toxic chemicals emitted into the air in the U.S. is around _____ metric tons annually.

63. At least _____ toxic pollutants have been identified.

64. List five standards included **maximum achievable control technology**.

a. _____

b. _____

c. _____

d. _____

e. _____

65.   MACT is expected to reduce emissions by at least _____ percent by the year 2005 at an estimated cost of $_____ billion.

Indoor Air Pollution

66.   Is it (True or False) that air inside the home and workplace often contains much higher levels of pollutants than outdoor air?

67.   Factors contributing to indoor air pollution include

a. products used indoors. (True or False)
b. well insulated and sealed buildings. (True or False)
c. duration of exposure to indoor pollutants. (True or False)

<u>**Sources of Indoor Pollution**</u>

68.   List five types or sources of indoor air pollutants. (see Figure 15-18)

a. _____   b. _____   c. _____

d. _____   e. _____

<u>**Public Policy on Indoor Pollution**</u>

69.   List the two sources of indoor pollution that EPA policies focus on.

a. _____   b. _____

Taking Stock
The Cost of Controlling Air Pollution

70.   Currently, pollution control costs are estimated at $_____ billion per year.

71.   Is it (True or False) that pollution control is now a major industry?

<u>**Future Directions**</u>

72.   List four future improvements that can be made to reduce air pollution.

a. _____   b. _____

c. _____   d. _____

Directions: Match each term with the list of definitions and examples given in the tables below. Term definitions and examples might be used more than once.

Term	Definition	Example (s)	Term	Definition	Example (s)
acute	j		point sources	p	
air pollutants	c		primary standard	n	
ambient standards	h		primary pollutants	l	
area sources	a		secondary pollutants	m	
carcinogenic	k		stratosphere	b	
catalytic converter	o		temperature inversion	g	
chronic	i		threshold level	d	
industrial smog	e		trophosphere	a	
photochemical smog	f				

VOCABULARY DEFINITIONS

a. lowest atmospheric layer

b. atmospheric layer immediately above the tropopause

c. substances in the atmosphere that have harmful effects

d. pollutant level below which no ill effects are observed

e. an irritating grayish mixture of soot, sulfurous compounds and water vapor

f. air pollutant resulting from sunlight interacting with primary sources

g. effect of certain weather conditions on smog levels

h. air pollutant levels that need to be achieved to protect environmental and human health

i. gradual deterioration of a variety of physiological functions over a period of years

j. life-threatening reactions within a period of hours or days

k. cancer-causing

l. direct products of combustion and evaporation

m. products from reactions between primary pollutants and atmospheric conditions

n. the highest level of pollutant that can be tolerated by humans without noticeable ill effects.

VOCABULARY DEFINITIONS
o. a device that removes toxic substances from internal combustion engines
p. direct pollution sources
q. diffuse pollution sources

VOCABULARY EXAMPLES	
a. Fig. 15-13	i. Fig. 15-4
b. bronchitis, fibrosis of the lungs	j. *Hard Times* by Charles Dickens
c. Fig. 15-17	k. household products
d. death	l. industry
e. Fig. 15-5	m. Table 15-1
f. Fig. 15-3	n. PM-10 for particulates
g. lungs	o. Fig. 15-1
h. Fig. 15-14	

SELF-TEST

1. Which of the following statements is not accurate concerning the air pollution problem?

    a. Organisms have the capacity to deal with certain amounts and levels of pollution.
    b. There is a threshold pollution level that is tolerable.
    c. Air pollution has been with human society since the discovery of fire.
    d. The solution to pollution is dilution.

2. Unprotected car exhaust + sunlight + a mountain topography =

    a. a temperature inversion b. photochemical smog c. an air pollution episode d. acid rain

3. The condition of a warm air mass overlying a cool air mass is referred to as

    a. a temperature inversion b. photochemical smog c. an air pollution episode d. acid rain

4. Air pollution may

    a. adversely affect human health.
    b. cause damage to crops and forests.
    c. increase the rate of corrosion and deterioration of materials.
    d. cause all of the above effects.

5.   A major problem with the Clean Air Act of 1970 and its 1977 and 1990 amendments is

       a. identifying pollutants.
       b. demonstrating cause and effect relationships.
       c. determining the source of pollutants.
       d. developing and implementing controls.

       Indicate whether the following air pollutants are Primary [P] or Secondary [S].

6.   [   ] particulates
7.   [   ] hydrocarbon emissions
8.   [   ] carbon monoxide
9.   [   ] nitric oxide
10.   [   ] sulfur dioxide
11.   [   ] ozone
12.   [   ] PANs
13.   [   ] VOC's
14.   [   ] lead
15.   [   ] sulfuric acid
16.   [   ] nitric acid

17.   A major change in the Clean Air Act of 1990 was:

       a. identifying criteria pollutants.
       b. imposition of sanctions on polluters.
       c. allowing polluters to choose the most cost-effective means of pollution control.
       d. all of the above.

18.   Air pollution is known to have adverse effects on

       a. human health   b. agricultural crops   c. forests   d. all of these

       List three effects of air pollution on human health.

19.   _____ 20. _____ 21._____

       Indicate whether the strategies described below were designed to control emissions of
       a. particulates, b. pollutants from motor vehicles, c. sulfur dioxide and acids,
       d. ozone, e. air toxics

22.   [   ] Phasing out the use of leaded gasoline.
23.   [   ] "No visible emissions"
24.   [   ] Putting catalytic converters on car exhaust systems.
25.   [   ] **"maximum achievable control technology"**
26.   [   ] Installation of filters and electrostatic precipitators at industrial sites.
27.   [   ] Tall smoke stacks
28.   [   ] Withholding federal highway funds.

List four indoor air pollutants

29. _____   30. _____

31. _____   32. _____

CHAPTER 16

MAJOR ATMOSPHERIC CHANGES

Chapter 16 focuses on the three pollution problems that threaten all life on Earth. The last chapter provided an overview of the variety of pollutants the human ecosystem expects the natural ecosystem to dilute, assimilate or to just make it go away! This chapter describes the peril and promise as the human ecosystem attempts to control acid precipitation, carbon dioxide emissions, and depletion of the ozone shield. All three problems, unless seriously addressed by citizens, local, state, and federal governments, will have permanent negative effects on the quality of life globally.

Acid precipitation alters the abiotic conditions of aquatic ecosystems so completely that the rich species diversity in lakes in the United States, Canada, and Europe is gone. Acid precipitation also alters nutrient cycling in terrestrial ecosystems leading to loss of plant life and erosion of topsoil. The sources of acid precipitation are well known. There is no doubt concerning cause-and-effect relationships! Slow or no responses by the government and industry have allowed continued and increased dumping of acid forming "oxides" into the atmosphere. It is easy, but regrettable, to make the predictions of expected outcomes if acid precipitation continues.

If you thought it was hot last summer, do not be surprised if global temperatures continue to rise and summer temperatures get hotter and hotter and hotter. Try to imagine the environmental and sociological impacts of melting polar ice caps, the bread basket of the world turning into a desert, and long-term droughts. These events have a high probability of occurring in your lifetime if reduced carbon dioxide emissions are not on the agendas of government and industry.

As if acid rain and the greenhouse effects were not enough to concern us, we must also contend with the potential of increased exposure to ultraviolet light - the ultimate suntan. The natural ecosystem provided a perfect shield (ozone) from ultraviolet light in the stratosphere. The human ecosystem has seen fit to destroy this life saving shield. Our actions regarding ozone are the ultimate paradox. We produce ozone at ground level as one component in photochemical pollution and destroy it in the place where it belongs and is of great benefit to us. The carcinogenic and mutagenic effects of ultraviolet light are well known. All life on Earth will be directly threatened by increasing ultraviolet radiation caused by depletion of the ozone shield.

There are solutions to the above three problems! In this chapter we will examine the causes of these three problems and what can be done to mitigate them.

STUDY QUESTIONS

Acid Deposition

1.   Industrialized regions of the world are regularly experiencing precipitation that is from 10 to _____ times more acid than normal.

2.   List four forms of acid precipitation (see Fig. 16-1).

a. _____   b. _____

c. _____   d. _____

Acids and Bases

3.   Indicate whether the following are properties of [a] acids, [b] bases, or [c] water.

[   ] Any chemical that will release OH^- ions.
[   ] Any chemical that will release H^+ ions.
[   ] The combination of one OH^- and 2 H^+ ions.
[   ] The balance or neutral point between acidic and basic solutions.

4.   **pH** is a measurement of (Hydrogen, Hydroxyl) ions in solution.

5.   On the pH scale, the number _____ represents the pH of pure water or neutral.

6.   On the pH scale, numbers decreasing from 7 (e.g., 6, 5, 4, 3, 2, 1) represent increasing (Acidity, Alkalinity).

7.   Each unit on the pH scale represents a factor of (1, 10, 100, 1000) in the acid concentration.

8.   A solution of pH 5 is (1, 10, 100, 1000) times more acid than a solution of pH 6.

9.   A solution of pH 4 is (1, 10, 100, 1000) times more acid than a solution of pH 6.

Extent and Potency of Acid Deposition

10.   Acid precipitation is defined as any precipitation with a pH of _____ or less.

11.   Give the pH values for the following examples.

a. precipitation in eastern U.S.: from _____ to _____
b. mountain forests east of Los Angeles: _____

Sources of Acid Deposition

12.   About two thirds of the acid in acid precipitation is _____ acid and one third is _____ acid.

13.   It is well known that sulfur and nitrogen are found in fossil _____.

14.   Indicate whether the following statements pertain to anthropogenic [a] or natural [n] oxide emissions.

[   ] produced by volcanic or lightening activity
[   ] produced by human activities
[   ] concentrated in industrial regions
[   ] spread out over the globe

15. In the Eastern U.S., the source of over 50% of acid deposition are _____ burning power plants.

16. Power plants attempt to alleviate sulfur dioxide emission at ground level by building (Shorter, Taller) stacks that ultimately dispersed the pollutant (Closer, Further) from the source.

Effects of Acid Deposition

<u>Impact on Aquatic Ecosystems</u>

17. pH controls all aspects of _____, _____, and _____.

18. Indicate whether the following conditions are [+] or are not [-] known effects of acid pH in aquatic ecosystems.

    [ ] Alteration of plant and animal reproduction.
    [ ] Introduction of other toxic elements (e.g. aluminum) from soil leachate.
    [ ] Shift from eutrophic to oligotrophic conditions.
    [ ] Totally barren and lifeless aquatic ecosystems.
    [ ] Alterations in the food chains.

19. How is it possible for some regions that receive roughly equal amounts of acid precipitation, do not also have acidified lakes?

    _____

20. Is it (True or False) that lakes can maintain their buffering capacity indefinitely?

21. Hydrogen ion concentrations in solution will (Increase, Decrease) in the presence of a buffer.

22. A mineral that is an important buffer in nature is _____, known chemically as calcium carbonate.

23. The amount of acid that a given amount of limestone can neutralize is known as its _____ capacity.

24. When an ecosystem loses its buffering capacity

    a. additional hydrogen ions will remain in solution. (True or False)
    b. the buffer is used up. (True or False)
    c. the ecosystem will become acidified. (True or False)

25. When the buffering capacity is exhausted, there is a (Rapid, Slow) drop in pH.

<u>Impact on Forests</u>

26. Indicate whether the below tree species are tolerant [T] or vulnerable [V] to acid rain.

    [   ] red spruce   [   ] balsam fir   [   ] sugar maple

27. Indicate whether the following are direct [D] or indirect [I] effects of acid precipitation on forest ecosystems (not covered in the text directly, but think about it!).

    [   ] Leaching of nutrients.
    [   ] Break down and release of aluminum into solution.
    [   ] Rapid changes in soil chemistry.
    [   ] Reduced growth and diebacks of plant and animal populations.
    [   ] Increased plant vulnerability to natural enemies and drought.
    [   ] Increased soil erosion.
    [   ] Increased flooding.
    [   ] Increased sedimentation of waterways.

Impact on Humans and Their Artifacts

28. Indicate whether the following effects of acid precipitation on humans is related to [a] artifacts, [b] health, or [c] aesthetics.

    [   ] mobilization of lead and other toxic elements
    [   ] corrosion of limestone and marble
    [   ] deterioration of lakes and forests

Coping With Acid Precipitation

29. Sulfur and nitrogen oxides are pumped into the troposphere at the rate of _____ million tons per year.

30. What is the source of acid deposition for the following countries?

    Canada: _____   Scandinavia: _____

    Japan: _____

Ways to Reduce Acid-Forming Emissions

31. How much of a reduction in acid-causing emissions is required to prevent further acidification of the environment? _____ %

    Identify the major problem(s) associated with the adoption of any of the following ways to reduce acid-forming emissions.

    [a] expense, [b] major reconstruction of industrial facility, [c] societal acceptance, [d] requires major shift in U.S. energy policy, [e] still have the problem of where to put the "waste"

32. fuel switching _____
33. coal washing _____
34. fluidized bed combustion _____
35. scrubbers _____
36. alternative power plants _____
37. reductions in electricity consumption _____

Indicate whether the following strategies for reducing acid-producing emissions are examples of [a] fuel switching, [b] coal washing, [c] fluidized bed combustion, [d] scrubbers, or [e] alternative power plants.

38.   [   ] Exhausting fumes through a spray of water containing lime.
39.   [   ] Combustion of coal in a mixture of sand and lime.
40.   [   ] Building more nuclear power plants.
41.   [   ] Pulverize and chemically wash coal.
42.   [   ] Using low-sulfur coals.

Political Developments

43.   The circumstantial evidence linking power plant emissions to acid deposition is (Insufficient, Overwhelming).

44.   Utility companies have (More, Less) governmental support than does the public on the acid precipitation issue.

45.   The federal government's response to the overwhelming evidence on the cause and effect relationship of acid deposition was more _____.

Title IV of the Clean Air Act of 1990

46.   What made Title IV a unique addition to The Clean Air Act?

        _____

47.   List the five provisions of Title IV.

        a. _____

        b. _____

        c. _____

        d. _____

        e. _____

Industry's Response to Title IV

48.   Which of the following have been industry's response to Title IV? (Check all that apply)

        [   ] fuel switching
        [   ] coal washing
        [   ] scrubbers
        [   ] emissions allowance trading
        [   ] using low-sulfur coals

Global Warming

The Earth as a Greenhouse

Warming Process

49.   Light energy is (Blocked, Absorbed) by greenhouse gases.

50.   Light energy is converted to _____ energy in the form of infrared radiation.

51.   When heat energy is trapped in an enclosed space and the temperature rises, this is known as the _____ effect.

52.   What experimental evidence demonstrates a 60% increase of atmospheric CO_2 since the ice ages?

        _____

Global Cooling

53.   Identify two factors that contribute to global cooling.

        a. _____   b. _____

The Carbon Dioxide Story

54. CO_2 levels are now _____ percent higher than before the Industrial Revolution.

55.   As carbon dioxide concentrations increase, global temperatures will (Increase, Decrease).

56.   A major *sink* for atmospheric CO_2 is the _____.

57.   Every kilogram of fossil fuel that is burned results in the production of an additional _____ kilograms of carbon dioxide.

58.   A total of _____ billion tons of carbon dioxide is added annually to the Earth's atmosphere from burning fossil fuels and tropical rain forests.

Other Greenhouse Gases

59.   Name and identify the sources of four other greenhouse gases

        Name   Source(s)

        _____   _____

        _____   _____

        _____   _____

        _____   _____

170

Amount of Warming and Its Probable Effects

60.   Identify four other variables that affect global temperatures.

   a. _____   b. _____

   c. _____   d. _____

The Contributions of Modeling

61.   If concentrations of greenhouse gases were to double, the earth would warm up between
        _____ and _____ degrees centigrade.

62.   Earth's temperatures were only _____ degrees C cooler during the ice age.

Impacts of Future Warming

63.   Two predicted major impacts of global warming are

   a. _____   b. _____

64.   Indicate whether the following are [+] or are not [-] probable effects of global warming.

   [ ] Melting of polar ice caps.
   [ ] Flooding of coastal areas.
   [ ] Massive migrations of people inland.
   [ ] Alteration of rainfall patterns.
   [ ] Deserts becoming farmland and farmland becoming deserts.
   [ ] Significant losses in crop yields.

Coping With Global Warming

Is Global Warming Here?

65.   List three sources of evidence that global warming might be here.

   a. _____   b. _____

   c. _____

66.   Is it (True or False) that carbon dioxide levels in the atmosphere are rising?

Energy Scenarios

67. CO_2 emissions are responsible for _____ percent of greenhouse warming.

Political Developments

68.　Which of the following are strategies to reduce carbon dioxide emissions and the effects of global warming? (Check all that apply)

[   ] limiting the use of fossil fuels in industry and transportation
[   ] adopt a wait-and-see attitude
[   ] develop and adopt alternative energy sources
[   ] encourage vast tree planting programs
[   ] examine other possible causes of global warming
[   ] make and enforce energy conservation rules
[   ] rely on the government to develop the needed strategies

Depletion of the Ozone Shield

Nature and Importance of the Shield

69.　Ultraviolet light is responsible for thousands of cases of _____ cancer per year in the U.S.

70.　_____ in the stratosphere prevents ultraviolet light from entering the Earth's atmosphere.

71.　The presence of this chemical barrier to ultraviolet light is known as the _____ shield.

72.　Is it (True or False) that the ozone in the stratosphere and the ozone that is a serious pollutant on the Earth's surface are the same ozone?

Formation and Breakdown of the Shield

73.　Complete the following reactions that form ozone (see Figure 16-15)

Reaction #1: _____ light + O2 → _____ + _____

Reaction #2: free _____ + O2 → _____

Reaction #3: free _____ + O3 → _____ + _____

Reaction #4: _____ light + O3 → _____ + _____

74.　Indicate whether O3 concentrations are high [+] or low [-] given the below conditions.

[   ] at the equator　　[   ] northern latitudes　[   ] winter　　[   ] summer

Halogens in the Atmosphere

75.　Identify four sources of CFCs.

a. _____　　b. _____

c. _____   d. _____

76.   A _____ is a chemical which promotes a chemical reaction without itself being used up by the reaction.

77.   Complete the following reactions that destroy ozone.

Reaction # 5: $CFCl_3$ + _____ light → _____ + $CFCl_2$

Reaction #6: Cl + _____ → _____ + O_2

Reaction #7: ClO + _____ → 2 _____ + O_2

78.   The source of chlorine entering the stratosphere is _____.

79.   Which of the above reactions releases Cl from CFCs (5, 6, 7)?

80.   Which of the above reactions generates more Cl (5, 6, 7)?

81.   Chlorine is **a catalyst** that destroys the product of reaction (1, 2, 3, 4)?

82.   Less ozone in the stratosphere will cause (More or Less) UV light penetration.

83.   Is it (True or False) that CFCs will eventually be flushed out of the atmosphere?

84.   Is it (True or False) that CFCs are water insoluble?

The Ozone "Hole"

85.   A gaping hole in the ozone shield was discovered over the _____ pole.

86.   This hole represented a _____ % reduction in ozone levels.

87.   The ozone hole:

a. persists throughout the year. (True or False)
b. reappears every year over the South Pole. (True or False)
c. is becoming larger at each appearance. (True or False)
d. may cause the destruction of marine phytoplankton. (True or False)
e. could be become more destructive if located over the North Pole. (True or False)

88.   Where is the ozone shield the thinnest? _____

89.   What is the predicted public health impact in this area?

_____

Controversy Over Volcanic Activity

90.   Is it (True or False) that volcanic eruptions inject significant amounts of Cl into the stratosphere?

91. Is it (True or False) that the Cl in the stratosphere from volcanic eruptions equals that of anthropogenic sources?

More Ozone Depletion

92. Indicate whether the following are [+] or are not [-] trends in the continuing saga of ozone depletion.

    a. [   ] high ClO over both poles   b. [   ] ozone holes over both poles
    c. [   ] ground-level UVB increases   d. [   ] skin cancer increases

93. Explain why ozone loss is expected to peak in 1998.

    _____

Coming to Grips With Ozone Depletion

94. Is it (True, False) that several nations and industries are scaling back CFCs production?

95. Define the **Montreal Protocol**:

    _____

96. Is it (True or False) that there are CFC substitutes for industrial use?

97. U.S. chemical companies must halt all CFC production by December 31, 19_____.

VOCABULARY DRILL					
Directions: Match each term with the list of definitions and examples given in the tables below. Term definitions and examples might be used more than once.					
Term	Definition	Example (s)	Term	Definition	Example (s)
acid precipitation			chlorine reservoirs		
acid deposition			chlorine cycle		
acids			chlorofluorocarbon		
anthropogenic			greenhouse gases		
artifacts			ozone shield		
bases			pH		
buffer			planetary albedo		
catalyst					

VOCABULARY DEFINITIONS

a. coming from human activities

b. precipitation that is more acidic than usual

c. concentration of H ions

d. an precipitation with a 5.5 pH or less

e. substance that has a large capacity to absorb H ions and hold pH relatively constant

f. human-made objects

g. atmospheric gases that play a role analogous to the glass in a greenhouse

h. a natural reflection of light that contributes to overall cooling

i. ozone in the stratosphere

j. organic molecules in which Cl and Fl replace some of the hydrogens

k. continual regeneration of Cl as it interacts with ozone

l. promotes a chemical reaction without itself being used up in the reaction

m. temporary reactions that hold Cl out of the chlorine cycle

n. H ion > OH ion concentration

o. H ion < Oh ion concentration

VOCABULARY EXAMPLES

a. $0 \rightarrow 7 \rightarrow 14$	h. limestone
b. clouds	i. monuments
c. CFC's, CO_2, CH_4	j. ph > 7
d. Cl	k. ph < 7
e. $ClO + ClO \rightarrow 2\,Cl + O_2$	l. Fig. 16-3
f. Fig. 16-1	m. NO_2, CH_4
g. halogenated hydrocarbons	n. Fig. 15-1

SELF-TEST

1. Any chemical that releases H^+ ions is

   a. an acid   b. a base   c. water   d. all of these

2.   Measurement of pH is actually a measure (hydrogen, hydroxyl) ions.

3.   A solution of pH 5 (1, 10, 100, 1000) times more acid than a solution of pH 6.

4.   About two thirds of the acid in acid precipitation is

      a. chromic   b. nitric   c. sulfuric   d. carboxylic acid

5.   It is well known that the source(s) of acid precipitation is (are)

      a. sulfur dioxide in coal.
      b. nitrogen oxides in gasoline.
      c. coal fired power plants.
      d. all of these

      Identify four direct or indirect effects of air pollutants on pine trees and the forest ecosystem in general.

6. _____

7. _____

8. _____

9. _____

      Identify the source and corresponding pollutant that cause each of the following environmental conditions.

              Sources   Pollutant

                            Acid Rain

      10. _____   11. _____

      12. _____   13. _____

                        Greenhouse gases

      14. _____   15. _____

      16. _____   17. _____

      18. _____   19. _____

      20. _____   21. _____

Depletion of
Ozone Shield

22. _____   23. _____

24. _____   25. _____

26. Define **buffer** and apply the term preventing acid deposition: _____

_____

Indicate whether the following environmental problems are or might be the result of [A] acid rain, [B] greenhouse effect, or [C] ozone depletion.

27. [   ] increased concentrations of UVB light hitting the earth's surface.
28. [   ] melting of polar ice caps.
29. [   ] deserts becoming farmland and farmland becoming deserts.
30. [   ] break-down and release of aluminum ions in solution.
31. [   ] higher incidence of skin cancer in humans.
32. [   ] shift of aquatic ecosystems from eutrophic to seemingly oligotrophic conditions.
33. [   ] flooding of coastal areas.

Identify whether the following pollution abatement procedures are [A] getting at the root cause or [B] attacking the symptoms.

34. _____ coal washing   35. _____ banning aerosol sprays   36. _____ using solar energy

37. _____ recapture CFCs   38. _____ increase fuel efficiency   39. _____ increase fuel costs

40. The ozone shield in the stratosphere protects Earth's biota from the damaging effects of

   a. CFCs   b. UV light   c. global warming   d. acid precipitation

CHAPTER 17

POLLUTION AND PUBLIC POLICY

At this point in your study of environmental science you should be developing an increasing awareness of the serious pollution problems that the humankind is imposing on natural ecosystems. Most water and air pollution problems have direct linkage with the economies and cultural priorities within developed countries. For example, enormous profits are realized by industries that produce the **things** demanded by the consumer public. People in developed countries have a profit mentality! Chapter 17 demonstrates how this profit mentality can serve to reduce rather than enhance water and air pollution. You are aware of several technologies that could be used to control water and air pollution but numerous competing , political, and economic interests often block action. This is the case because industries may view the implementation of these technologies as unnecessary overhead costs that would reduce their profit margin. In fact, it is good business to control pollution. Imagine the amount of money lost to industries and stock holders when employees are too sick to work, health insurance costs rise, equipment deteriorates, and when the costs of raw materials rise because of limited supplies. What are the long-term risks associated with continued water and air pollution compared to the economic benefits that could be derived from the utilization of pollution abatement technologies? The objective of Chapter 17 is to examine this question.

STUDY QUESTIONS

Origins of Environmental Public Policy
The Need for Environmental Public Policy

1.   Identify the two areas of emphasis in promoting the common good with environmental public policies.

      a. _____

      b. _____

2.   What is the message portrayed in Table 17-1 concerning the need for environmental public policy?

      _____

Relationships Between Economic Development and the Environment

3.   Examine Fig. 17-2 and determine whether the relationships between income and environmental variable are [D] direct, [I] inverse, or both [B].

      [   ] population without safe water
      [   ] population without adequate sanitation
      [   ] urban concentrations of SO2
      [   ] municipal wastes per capita
      [   ] CO2 emissions per capita

4.   Match the below list of policy life cycle characteristics with the four stages: [a] recognition, [b] formulation, [c] implementation, or [d] control.

[   ] real political and economic costs of a policy are exacted
[   ] low in political weight
[   ] policies broadly supported
[   ] media has popularized the policy
[   ] rapidly increasing political weight
[   ] dissension is high
[   ] public concern and political weight are declining
[   ] policies broadly supported
[   ] media coverage is high
[   ] issue not very interesting to media
[   ] environment is improving
[   ] debate about policy options occurs

5.   Using the same stages as listed above, indicate the status of policies that address the following environmental problems (see Fig. 17-4).

[   ] toxic chemicals
[   ] sewage water treatment
[   ] global warming
[   ] ozone depletion
[   ] urban sprawl

Economic Effects of Environmental Public Policy

6.   The best environmental policies are those that conform to the _____, _____, and _____ criteria.

Costs of Policies

7.   Is it (True or False) that some policies have no direct monetary costs?

8.   Is it (True or False) that some policies have no political costs?

Impact on the Economy

9.   Is the relationship between strict environmental regulations and economic growth (Positive or Negative)?

10.   Summarize the general conclusions of studies on the impact of environmental policy on the economy.

_____

Implementing Environmental Public Policy
Adopting Good Policies

11. _____ and _____ are essential for formulating good environmental policies.

12. Evaluate the following environmental policies in terms of benefits and costs (see Fig. 17-6).

    Taxes on urban road congestion: _____

    _____

    Taxes on emissions and wastes: _____

    _____

    Investments in water supply and sanitation: _____

    _____

Policy Options: Market or Regulatory?

13. Indicate whether the following statements or policy examples characterize the market [M] or regulatory [R] approaches.

    [   ] Clean Air Act of 1990
    [   ] the most common approach in environmental public policy
    [   ] practically guarantees a certain sustained level of pollution

Cost-Benefit Analysis

14. Explain why Executive Order 12291 was issued by President Reagan.

    _____

15. A comparison of costs and benefits is commonly called a _____ **analysis**.

16. **Cost-effective** projects have benefits that are (Greater, Less) than the costs.

17. The effect of the business process not included in the usual calculations of profit and loss is called an _____.

18. Give an example of a:

    a. good externality _____

    b. bad externality _____

The Costs of Environmental Regulations

19.   Is it (True or False) that most forms of pollution control involve additional expenses?

20.   List two ways that pollution control increases costs.

     a. _____

     b. _____

21.   Is it (True or False) that 100% pollution control is cost-effective? (see Fig. 17-7a)

22.   Costs of pollution control are the highest in the (Early, Later) stages of control (see Fig. 17-8a).

The Benefits of Environmental Regulation

23.   List six major benefits derived from pollution control (see Table 17-2).

     a. _____

     b. _____

     c. _____

     d. _____

     e. _____

     f. _____

24.   Give one method of estimating benefits derived from pollution control.

     _____

25.   At what level of pollution control will the benefits be negligible and not worth the cost?
     _____ % or below _____ level

Cost Effectiveness

26.   Optimum cost-effectiveness is achieved when the benefit curve is the (Greatest, Least) distance above the cost curve (see Figure 17-7c).

27.   Is it (True, False) that equipment, labor, and maintenance costs can be estimated with a fair degree of objectivity?

28.   Cost of pollution controls will be (High, Low) at the time they are initiated and tend to (Increase, Decrease) as time passes. (see Figure 17-7a)

29. Indicate whether the following benefits or risks can be assigned a monetary value through [a] real dollar estimates, [b] **shadow pricing**, or [c] cannot be given a value.

[   ] Health benefits from eliminating air pollution episodes.
[   ] Costs of maintaining and replacing materials.
[   ] Income from water recreation.
[   ] Improving air quality.
[   ] Impact on nonhuman environmental components.
[   ] Assessing the value of human life.

Problems in Comparing Costs and Benefits

30. Is it (True or False) that some situations that may appear to be cost-ineffective in the short term may be extremely cost-effective in the long term?

31. Identify two types of pollution that may apply to a "true" response to the previous question.

a. _____   b. _____

32. Is it (True or False) that those who bear the cost of pollution control and those who benefit from these controls are different groups of people?

33. Give one example of a situation that would support a "true" response to the previous question.

_____

Progress and Public Support

34. What percent of the U.S. public:

a. are convinced that pollution affects the quality of their lives? _____
b. believe that current environmental laws are too weak? _____
c. believe that environmental improvements should be made regardless of cost? _____

35. Indicate whether the following have [+] or have not [-] been EPA achievements from 1970-1990.

[   ] reduction in vehicle emissions
[   ] construction of hundreds of wastewater treatment facilities
[   ] eliminating ocean dumping
[   ] stopped land disposal of untreated hazardous wastes
[   ] cleaned up 52 hazardous waste sites
[   ] banned production and use of asbestos, DDT, PCBs and leaded gasoline

36. The phase out of leaded gasoline cost industry $ _____ billion but accrued benefits represented $ _____ billion.

Risk Analysis

37.   Which of the following actions carry the greatest risk? (see Table 17-3)

a. being a policeman   b. drinking alcohol   c. smoking cigarettes

Risk Analysis by the EPA

38.   Risk analysis includes

a. hazard assessment. (True or False)
b. dose-response assessment. (True or False)
c. exposure assessment. (True or False)
d. risk characterization. (True or False)

39.   **Epidemiological studies** are used to make a cause and effect relationship between human health and (Past, Future) exposures to carcinogenic chemicals.

40.   **Animal tests** are used to make a cause and effect relationship between human health and (Past, Future) exposures to carcinogenic chemicals.

41.   Match the steps in risk analysis with the methods used to perform each step.

Step   Method

[ ] hazard assessment   a. final estimate of probability of fatal outcome from hazard
[ ] dose-response assessment   b. associates dose with incidence and severity of response
[ ] exposure assessment   c. uses epidemiological studies and animal tests
[ ] risk characterization   d. studies human groups with previous exposures

Risk Perception

42.   How many of the public's top 11 environmental concerns are also on the EPA's list of top 11 risks (see Table 17-5). _____

43.   Is the public's **risk perception** (the Same, Different) from that of EPA scientists?

44.   List the outrage factors that influence the public's risk perception.

a. _____   b. _____   c. _____

d. _____   e. _____   f. _____

g. _____

45.   Is it (Public Concern or Cost-benefit Analysis) the drives public policy?

46.   Is it (True or False) that public concern for human health overrides concern for major ecological risks.

Risk Management

47.   List the two steps in **risk management**.

    a. _____

    b. _____

48.   Regulatory decisions are based on:

    a. cost-benefit analysis   b. risk-benefit analysis   c. public preferences   d. all of these

The Public and Public Policy

49.   Match the below types of public involvement in environmental policy development with the [a] recognition, [b] formulation, [c] implementation, and [d] control stages in the policy life cycle.

    [ ] grassroots concern about environmental problems
    [ ] informing constituencies about the problem
    [ ] informing elected officials about level of public support for environmental policy
    [ ] electing or defeating political candidates
    [ ] public comment on regulations

		VOCABULARY DRILL			
Directions:		Match each term with the list of definitions and examples given in the tables below. Term definitions and examples might be used more than once.			
Term	**Definition**	**Example (s)**	**Term**	**Definition**	**Example (s)**
benefit-cost			market approach		
cost-benefit analysis			outrage		
cost-effective			policy life cycle		
dose			regulatory approach		
dose-response			risk characterization		
ecological risks			risk		
exposure			risk analysis		
externality			risk perceptions		
hazard			risk management		
health risks			shadow pricing		

VOCABULARY DEFINITIONS
a. recognition, formulation, implementation, control
b. set prices on pollution and resource use
c. standards are set and technologies prescribed
d. compares the estimated costs of project with the benefits that will be achieved
e. all cost and benefits are given monetary costs and compared
f. economic justification for proceeding with a given project
g. an effect of the business process not included in calculations of profit and loss
h. asking people what they might pay for a particular benefit
i. anything that can cause injury, damage, or deterioration
j. probability of suffering from injury, disease, or death
k. process of evaluating the risks associated with a particular hazard before taking the action
l. concentrations of the chemical in a test
m. incidence and severity of response when exposed to different doses of a chemical
n. identifying subjects, origin of chemical, dose and length of exposure
o. probability of a fatal outcome due to a hazard
p. concerning the public (human) welfare
q. concerning nature's welfare
r. intuitive judgement about risks
s. additional public concerns other than fatalities
t. risk characterization of a hazard followed by regulatory decisions

VOCABULARY EXAMPLES	
a. a favorable cost-benefit ratio	i. involuntary risks
b. Fig. 17-7	j. Fig. 17-3
c. Table 17-3	k. toxic chemicals and cancer
d. command-and-control strategy	l. Table 17-5
e. trading of emissions allowances	m. willingness to pay
f. How many will die?	n. Table 17-4
g. Chernobyl, Russia	o. workers improve job performance
h. Fig. 17-6	

SELF-TEST

1. Order (1 to 4) the stages of the policy life cycle..

    [   ] implementation   [   ] recognition   [   ] control   [   ] formulation

2. What are the two costs in development of environmental policies?

    a. _____   b. _____

3. Is it (True or False) that a strong set of environmental policies means diminished national wealth?

4. List two areas of emphasis in environmental policy development.

    a. _____   b. _____

5. Using Fig. 17-7, explain the cost of pollution control over time given the percent of pollution reduction.

    _____

6. At what level (%) of pollution control are benefits maximized? _____

7. How does the Fig. 17-7 demonstrate that 100% reduction in pollution is not cost-effective?

    _____

8. Identify the area between the two curves in Fig. 17-7 that represents the area of optimum cost-effectiveness.

    _____

    Indicate whether the following benefits from reduction or prevention of pollution can be assigned a monetary value through [a] real dollar estimates or [b] shadow pricing.

9.   [   ] improved human health
10. [   ] improved agriculture and forest production
11. [   ] enhanced commercial and/or sport fishing
12. [   ] enhancement of recreational opportunities
13. [   ] extended lifetime of materials and cleaning
14. [   ] enhanced real-estate values

15. The costs over time for pollution control

    a. are greatest in the short-term and diminish in the long-term.
    b. are low in the short-term but increase exponentially the longer the control is in effect.
    c. are minimal at first, then increase to a maximum, then decrease again
    d. are maximal at first, then decrease to a minimum, then increase again

16.   Using Fig. 17-8, describe what happens to pollution control costs each year after time of initiation.

_____

17.   At what point in time, does Fig. 17-8 demonstrate a favorable cost-benefit ratio. _____ years.

Match the below definitions [a] hazard, [b] risk, [c] outrage

18.   [   ] People who have a choice will accept risk at a greater rate than having no choice.
19.   [   ] The probability of suffering injury or other loss as a result of exposure to hazard.
20.   [   ] Anything that has the potential of causing suffering due to exposure.

Match the steps in risk analysis with the methods used to perform each step.

Step	Method
21. [ ] hazard assessment	a. final estimate of probability of fatal outcome from hazard
22. [ ] dose-response assessment	b. associates dose with incidence and severity of response
23. [ ] exposure assessment	c. uses epidemiological studies and animal tests
24. [ ] risk characterization	d. studies human groups with previous exposures

CHAPTER 18

WILD SPECIES: BIODIVERSITY AND PROTECTION

How many rivets do you think there are in a Boeing 747? There are probably thousands but each single rivet is as important as the thousands in terms of maintaining the structural integrity of the entire airplane. Now imagine the following series of events take place on your next trip. You have just boarded the plane and are comfortably settled into your seat. You look out the window and see a person on the wing. This person is taking one of the rivets out of the wing! You are (or should be) alarmed and call the flight attendant. You complain that it is not a good thing for the person to be taking a rivet out of the wing of the plane. The flight attendant calms your fears by telling you that you should not worry because there are thousands of rivets in the wing of this plane. She further tells you that the person extracting the rivets is getting $5.00 per rivet and will be extracting one rivet from the other wing also. Well, you take off and arrive safely at your destination. On your return flight, the same thing happens but now the person is extracting two rivets from each wing. The stewardess tells you that again it is alright because the person is now getting $10.00 per rivet. Well this story could go on and on but let's get to the point. How many rivets, regardless of the economic value, would you allow to be taken out of the plane before you fear for its structural integrity?

Let us ask the question in a manner that is more relevant to the conceptual framework of Chapter 18 How many species (rivets) will you allow to be eliminated from an ecosystem (Boeing 747) before you fear for its structural integrity? Previous chapters have stressed the need to maintain ecosystem sustainability. Each species of plant and animal plays a vital role in this process even though we may have not yet identified the importance of every single species. Nevertheless, we are rapidly pulling out the rivets that hold the ecosystem together through over harvest, destruction of habitats, and major alterations in the abiotic characteristics of ecosystems through pollution. Chapter 18 examines the values of natural biota, how the human ecosystem is sacrificing these values, and what needs to be done to preserve natural biota.

STUDY QUESTIONS

Saving Wild Species
Game Animals in the United States

1. Was it (Commercial or Regulated) hunting that caused the extinction or near extinction of some game animals?

2. Restrictions on hunting that represent wildlife management are

    a. requiring a hunting license. (True or False)
    b. limiting the type of weapon used in the harvest. (True or False)
    c. hunting within specified seasons. (True or False)
    d. setting size and/or sex limits on harvested animals. (True, False)
    e. limiting the number that can be harvested. (True or False)
    f. prohibiting commercial hunting. (True or False)

3.   Which of the following do [+] and do not [-] represent contemporary wildlife problems.

        [   ] near extinction of the wild turkey
        [   ] road killed animals
        [   ] urban wildlife, e.g., opossums, skunks, raccoons
        [   ] lack of natural predators
        [   ] wildlife as vectors for certain diseases

The Endangered Species Act

4.   Snowy egrets were heavily exploited in the 1800s for their (a. meat b. eggs c. nests d. feathers) that were used in (a. fancy restaurants b. ladies hats c. curiosity shops d. grocery stores).

5.   The two states that were first to pass laws protecting plumed birds were _____ and _____.

6.   The law that forbids interstate commerce in illegally killed wildlife is called the _____ Act.

7.   The law that protects species from extinction is called the _____ _____ Act of _____.

8.   Indicate whether the following provisions are strengths [+] or weaknesses [-] of the Endangered Species Act.

        [—] The need for *official recognition*.
        [+] Stiff fines for the killing, trapping, uprooting, or commerce of endangered species.
        [+] Controls on government development projects in critical habitats.
        [—] Enforcement of the act.
        [—] recovery programs
        [—] taxonomic status of some animals protected by the act

9.   Identify two endangered bird species that were "recovered".

        a. _____   b. _____

10.   Explain the connection between sandhill and whooping cranes.

        _____

11.   What major concerns of the general public may scuttle the reauthorization of the ESA in the future? property rights and loss of jobs

        _____

Value of Wild Species
Biological Wealth

12.   We presently know of the existence of _____ million species and estimate that at least _____ to _____ million additional species have not been studied.

13.   How many plant and animal species have become extinct in the United States? _____

Instrumental Versus Intrinsic Value

14.   List the five values of natural species.

a. _____

b. _____

c. _____

d. _____

e. _____

Sources for Agriculture, Forestry, Aquaculture, and Animal Husbandry

15.   Indicate whether the following statements pertain to [a] wild populations of plants or animals or [b] cultivated (**cultivar**) populations of plants or animals.

[   ] Highly adaptable to changing environmental conditions.
[   ] Highly nonadaptable to changing environmental conditions.
[   ] Have numerous traits for resistance to parasites.
[   ] Lack genetic **vigor**.
[   ] Can only survive under highly controlled environmental conditions.
[   ] Have a high degree of genetic diversity in the gene pool of the population.
[   ] Have a low degree of genetic diversity in the gene pool of the population.
[   ] Represents a reservoir of genetic material commonly called a **genetic bank**.

16.   Of the estimated 7,000 plant species existing in nature, how many have been used in agriculture? _____

17.   What is the significance of the winged bean (Fig. 18-4) in regard to the answer to the previous question?

_____

Sources for Medicine

18.   Of what value is the chemical called vincristine?

_____

19.   Vincristine is an extract from the _____

20.   Of what value is the chemical called capoten?

_____

21.   Capoten is an extract from the _____

22.	Is it (True or False) that there is the potential of discovering innumerable drugs in natural plants, animals, and microbes?

Commercial Value

23.	One of the largest income-generating enterprises in many developing countries is

_____

24.	As leisure time (Increases, Decreases), larger portions of the economy become connected to supporting activities related to the natural environment.

25.	Pollution (Increases, Decreases) the commercial benefits of the natural environment.

26.	The contribution of wild plants and animal species to the U.S. economy was calculated to be more than $ _____ billion.

✓ 27.	Commercial interests often result in the (Conservation, Destruction) of natural biota.

Recreational, Aesthetic, and Scientific Value

28.	Indicate whether the following are [a] recreational, [b] aesthetic, or [c] scientific values of the biota of natural ecosystems.

   [   ] hunting
   [   ] sport fishing
   [   ] hiking
   [   ] camping
   [   ] Being in a forest rather than a city dump.
   [   ] study of ecology
   [   ] Just knowing that certain plant and animal species are alive.

Intrinsic Value

29.	Intrinsic value arguments center on _____ rights and _____ .

Biodiversity

30.	Over geological time, the net balance between _____ and _____ has favored the gradual (Accumulation, Loss) of species.

31.	The number of plant and animal species on the earth today is estimated at _____ million with the greatest species richness found in _____ plants and _____ .

32.	Which country has at least 5% of all living species? _____

The Decline of Biodiversity
Losses Known and Unknown

How many wild plant and animal species in the United States

33.   are at risk of being lost? _____
34.   have become extinct since the days of colonization? _____
35.   are on the endangered or threatened lists? _____

36.   In what type of ecosystem can one find the greatest biodiversity? _____

37.   List four sources of evidence that there is a decline of biodiversity occurring.

        a. _____   b. _____

        c. _____   d. _____

38.   Give two reasons why species on oceanic islands are more susceptible to extinction than
        continental species.

        a. _____

        b. _____

Reasons for the Decline

39.   List five factors that represent evidence of or contribute to the decline in biodiversity.

        a. _____

        b. _____

        c. _____

        d. _____

        e. _____

        Match the following examples of factors the represent evidence of or contribute to the decline of
        biodiversity with those you listed [a through e] in the previous question.

40.   [   ] habitat fragmentation
41.   [   ] clear cut logging
42.   [   ] expansion of the human population over the globe
43.   [   ] the brown tree snake
44.   [   ] the greenhouse effect
45.   [   ] Table 18-1
46.   [   ] prospects of huge immediate profits
47.   [   ] the growing fad for exotic pets, fish, reptiles, birds, and house plants

Physical Alteration of Habitats

48.   Indicate whether the following examples of habitat alteration represent [a] conversion, [b] fragmentation, or [c] simplification

[   ] stream channelization
[   ] shopping malls
[   ] highway development

The Population Factor

49.   What is the graphic relationship (see Figure 18-9) between human population size and species survival worldwide? _____

50.   As human population increases, species survival worldwide will (Increase, Decrease).

Pollution

51.   The greatest catastrophe to hit natural biota in 60 million years will be

a. acid rain   b. human population growth   c. global warming   d. urbanization

52.   Species adapt very (Quickly, Slowly) to environmental change.

53.   Global warming is predicted to occur very (Quickly, Slowly).

Exotic Species

54.   Exotic species are responsible for _____ percent of all animal extinctions since 1600.

55.   Usually, the introduction of exotic species into a country (Increases, Decreases) biodiversity.

56.   Usually, an exotic species as a (More, Less) effective competitor for resources.

57.   Give three examples of exotic species introduced into the United States.

a. _____   b. _____   c. _____

Overuse

58.   As certain plants and animals become increasingly rare, the price people are willing to pay will (Increase, Decrease).

59.   Would you predict that exploiters of natural biota would (Encourage, Discourage) public education on poaching and black market trade?

60.   Below is a chain of events that perpetuates overuse of natural biota. Which link in the chain needs to be removed to most effectively curtail overuse?

a. hunter b. trading post c. taxidermist d. tourist shop e. consumer tourist

61.   Give an example of human use of the following wildlife or plant species or their parts..

Small songbirds: _____

Tropical hardwoods: _____

Bear gall bladders: _____

Parrots: _____

Consequences of Losing Biodiversity

62.   Explain why **keystone species** are such an important group to consider as a consequence of losing biodiversity?

_____

International Steps to Protect Biodiversity

63.   List two international steps that have been taken to protect biodiversity.

a. _____   b. _____

64.   Which of the above steps focuses on

a. trade in wildlife and wildlife parts? _____

b. conserving biological diversity worldwide? _____

c. does not yet have political support in the U.S.? _____

Directions: Match each term with the list of definitions and examples given in the tables below. Term definitions and examples might be used more than once.

Term	Definition	Example (s)	Term	Definition	Example (s)
anthropocentric			fragmentation		
biodiversity			genetic bank		
biological wealth			instrumental value		
biota			intrinsic value		
conversion			keystone species		
cultivar			simplification		
ecotourism			speciation		
endangered species			threatened species		
exotic species			vigor		
extinction					

VOCABULARY DEFINITIONS

a. reduced to a point where it is eminent danger of extinction

b. judged to be in jeopardy but not on brink of extinction

c. assemblage of all living organisms

d. assemblage of all living organisms and their ecosystems

e. existence benefits some other entity

f. an entity that derives instrumental value from biota

g. something that has value for its own sake

h. tolerance to adverse conditions

i. cultivated variety of a wild strain

j. source of all genetic traits of wild strains of plants and animals

k. visiting a place in order to observe unique ecological sites

l. the creation of new species

m. the disappearance of species

n. human development of natural areas

o. process of cutting natural areas into separated parcels

VOCABULARY DEFINITIONS
p. diminishing abiotic and biotic habitat diversity
q. species introduced into an area from somewhere else
r. a species whose role is absolutely vital for the survival of many other species in the ecosystem

VOCABULARY EXAMPLES	
a. 1.4 million species	i. housing subdivisions
b. channelization	j. humans
c. competitiveness	k. Table 18-1
d. corn, wheat, rice	l. Fig. 18-2
e. Darwin's finches	m. large predators
f. Fig. 18-3	n. cats
g. Galapagos Islands	o. sources for agriculture
h. highways	p. just knowing that it exists

SELF-TEST

Indicate whether the following species have basically [a] instrumental or [b] intrinsic value.

1.   [   ] tuna
2.   [   ] butterflies
3.   [   ] rosy periwinkle
4.   [   ] hummingbirds

Match the examples on the left with the values of natural species on the right.

Examples   Values

5.   [   ] cultivars   a. intrinsic
6.   [   ] captone   b. commercial
7.   [   ] bird watching   c. recreational
8.   [   ] ecotourism   d. medicinal
9.   [   ] just knowing the species exists   e. agricultural

10.   The intrinsic value of wild species is rooted in _____ rights and _____ beliefs.

11.   Biodiversity is measured by the _____ of different species in an ecosystem.

12.   Which of the following types of ecosystems has the greatest biodiversity?

       a. grassland   b. tropical rain forest   c. urban development   d. cornfield

       Indicate whether the following activities conserve [+] or destroy [-] biodiversity.

13.   [   ] conversion of habitat
14.   [   ] pollution
15.   [   ] introduction of exotics
16.   [   ] expansion of human population globally
17.   [   ] overuse
18.   [   ] regulated hunting
19.   [   ] recovery plans
20.   [   ] ecosystem management

       Complete the statements 21 and 22 using the terms MORE or LESS.

21.   More expansion of the human population leads to _____ pollution, _____ use of
       wildlife species, and _____ biodiversity.

22.   More introduction of exotics leads to _____ competition for resources, _____
       predation on endemic prey species, and _____ biodiversity.

23.   The consequence(s) of species extinctions is (are):

       a. cascading loss of other species.
       b. loss of nature's services.
       c. loss of genetic banks.
       d. all of the above.

24.   Was it (Commercial, Regulated) hunting that caused the extinction or near extinction of some
       wild animals?

       Indicate whether the following statements pertain to [a] ESA, [b] CITES, [c] Convention on
       Biological Diversity, or [d] more than one of these.

25.   [   ] focuses on the trade of wildlife or their parts
26.   [   ] uses recovery plans to re-establish species numbers
27.   [   ] emphasizes sovereignty of a nation over its biodiversity

CHAPTER 19

ECOSYSTEMS AS RESOURCES

It takes years of training and experience to learn how to "manage" a single species. Try to imagine what intellectual skills must be brought to bear on whole ecosystem management, how many players (e.g., society, industry, politicians) participate in the management process, and how many personal agendas are involved? For example, ecosystem management may dictate that wolves be returned to the Yellowstone National Park or that cougars be protected in rangeland areas. Your mind must leap to the conclusion that some subsets of society will vigorously favor and fight such a management decision. Put simply, successful ecosystem management will depend more on public understanding of ecosystem structure and function than a highly trained wildlife or fisheries biologist. We have a long way to go in public education on ecosystems since the majority of people probably still believe that heat comes from a furnace and food from a grocery store. Chapter 19 provides the background information on ecosystem management and gives many applications. This chapter also reinforces the need for public understanding of ecosystem management concepts given the consequences of the lack of this understanding.

STUDY QUESTIONS

Ecosystems Under Pressure

1. List four ecosystems that are under pressure.

    a. _____   b. _____

    c. _____   d. _____

Forests and Woodlands

2. The major threat to the world's forests is total _____.

3. Worldwide, (a. 3/3, b. 2/3, c. 1/3) of the area originally covered by forests is now devoid of trees.

4. Indicate whether clearing forests increases [+] or decreases [-]:

    [   ] productivity
    [   ] nutrient recycling
    [   ] biodiversity
    [   ] soil erosion
    [   ] transpiration

5. List five alternatives to clear-cut logging.

    a. _____

b. _____

c. _____

d. _____

e. _____

Tropical Forests

6. Tropical forest are being destroyed at a rate of over _____ million acres/year.

7. Tropical rainforests provide

   a. millions of plant and animal species. (True or False)
   b. a system that maintains the climate of the earth. (True or False)
   c. a system that controls air pollution. (True or False)

8. List the two basic causes of deforestation.

   a. _____   b. _____

9. Indicate whether the following conditions promote [+] or prevent [-] the destruction of the tropical rainforest.

   [   ] colonization of forested lands
   [   ] huge national debts
   [   ] fast food chains and cheap hamburger
   [   ] loans from the World Bank
   [   ] **extractive reserves**
   [   ] environmental impact assessments
   [   ] **Statement on Forest Principles**

Oceanic Ecosystems
Marine Fisheries

10. Open oceans were traditionally considered an _____ commons.

11. As a response to over fishing in the international commons, many nations extended their _____ limits.

12. In 1977, the United States extended their territorial limits from 3 to 12 miles to _____ miles offshore.

13. Oceans could yield at least _____ million metric tons on a sustained basis.

14. What is the best strategy to restore a fishery like George's Bank?

   _____

International Whaling

15. Indicate whether the following groups or events have had a [+] or [-] effect on whale populations.

    [   ] International Whaling Commission
    [   ] International Union for Conservation of Nature
    [   ] whale watching
    [   ] Japan's harvest of whales for *scientific research*
    [   ] Stellwagen Bank

16. What three countries want whaling to continue?

    a. _____   b. _____

    c. _____

17. Why do the above three countries want whaling to continue?

    _____

18. Today, (Pollution, Over harvest) is the greatest threat to freshwater and estuarine fisheries.

Coral Reefs and Mangroves

19. Indicate whether the below sources of damage impact coral reefs [C] or mangroves [M].

    [   ] lucrative tropical fish trade
    [   ] shrimp aquaculture
    [   ] islander poverty
    [   ] logging

20. Indicate whether the below statements represent values of coral reefs [C] or mangroves [M] or both [B]

    [   ] protect coasts from storm damage and erosion
    [   ] form a rich refuge and nursery for many marine fish
    [   ] important food source for local people

Biological Systems in a Global Perspective
Major Systems and Their Services

21. Identify the largest (% of total, Table 19-2) and smallest terrestrial ecosystems.

    Largest = _____   Smallest = _____

22. List eight services performed by natural ecosystems.

    a. _____

b. _____

c. _____

d. _____

e. _____

f. _____

g. _____

h. _____

23. The loss of which natural service led to:

    a. siltation and flooding of rivers in Bangladesh? _____

    b. eutrophication of Chesapeake Bay? _____

24. How much would it cost to duplicate the water purification and fish propagation capacity of a single acre of tidal wetland? $ _____ per year

25. Is it (True, False) that we can compensate for the losses of natural services?

Ecosystems and Natural Resources

26. In what two ways does society assign value to natural resources?

    a. _____   b. _____

27. The term **natural resource** more often than not is defined in an (Ecological, Economic) setting.

Environmental Accounting

28. What value has traditionally been left out of the calculation of GNP?

    _____

29. Now, GNP accounting ledgers show the disappearance of millions of acres of forest as an economic (Asset, Loss).

30. Environmental accounting will generally (Increase, Decrease) GNP figures.

Conservation and Preservation

31. Any resource that has the capacity to renew or replenish itself is called _____ **resource**.

32. Is it (True, False) that **conservation** implies complete denial of any use of a natural resource?

33.   The aim of conservation is to _____ or _____ use.

34.   Is it (True, False) the **preservation** implies complete denial of any use of a natural resource?

Patterns of Use of Natural Ecosystems
Tragedy of the Commons

35.   Which of the following actions or statements are [+] or are not [-] characteristic of the *tragedy of the commons*?

[   ] Whoever grazes the most cattle gains an economic advantage over those who graze less.
[   ] "If I don't harvest this resource, someone else will."
[   ] Reacting to diminished catches of clams by harvesting less.
[   ] Whenever two or more independent groups are engaged in exploitation of a resource.
[   ] Eventually no one will be grazing cattle or digging clams.

36.   The tragedy of the commons can be avoided by limiting _____.

37.   What are two strategies for limiting access to the commons.

       a. _____   b. _____

38.   List the conditions for regulated access to the commons.

       a. _____

       b. _____

       c. _____

Maximum Sustainable Yield

39.   Match the following definitions with [a] maximum sustained yield and [b] carrying capacity.

[   ] The maximum population that an ecosystem can support.
[   ] The maximum use a system can sustain without impairing its ability to renew itself.

40.   Maximum sustained yield is obtained with (Optimal, Maximum) population size.

41.   Maximum sustained yield is (Reached, Exceeded) when use begins to destroy regenerative capacity.

42.   Indicate whether the following population characteristics will increase [+] or decrease [-] sustained yield.

[   ] A population below or within the carrying capacity.
[   ] A population that is approaching or exceeding carrying capacity.

43.   Is it (True or False) that carrying capacity and optimal population size are constant?

44.   Indicate whether the following procedures will increase [+] or decrease [-] sustained yield.

[   ] Pollution and other forms of habitat alteration.
[   ] Use levels that ignore population characteristics in relation to carrying capacity.
[   ] Using suitable management procedures.
[   ] Harvesting populations that are at or exceeding the carrying capacity.

45.   Indicate whether the following procedures will sustain [+] or diminish [-] maximum sustained yield.

[   ] Protecting biota from the "tragedy of the commons."
[   ] Invoking regulations but without enforcement.
[   ] Reducing the economic incentives that promote violation of regulations.
[   ] Preserving habitats.
[   ] Protecting habitats from pollution.

Restoration

46.   What is the intent of **restoration ecology**?

_____

47.   What is the primary requirement in restoration ecology?

_____

48.   Give four examples of ecological problems that can be ameliorated by restoration.

a. _____   b. _____

c. _____   d. _____

Public Lands in the United States

49.   Nearly _____ percent of the U.S. land is publicly owned.

50.   The majority of U.S. public land are in the _____ states and _____.

51.   Is the purpose of the Wilderness Act of 1964 to (Conserve, Preserve)?

52.   Identify the federal agencies that manage:

a. National Parks. _____

b. National Wildlife Refuges. _____

c. National Forests. _____ and _____

National Forests

53. Three-fourths of the managed commercial forests are (Private, State, Federal) lands.

54. Is it (True or False) that deforestation is no longer a problem in the U.S.?

55. Only _____ percent of the original U.S. forests are left.

56. Most U.S. forests are (Old, Second) - growth forests.

57. Is it (True or False) that U.S. taxpayers make money from the sale of national forest lumber?

58. List four factors that characterize "New Forestry."

    a. _____

    b. _____

    c. _____

    d. _____

Wise Use - Environmental Backlash

59. What appears to be the agenda of "Wise Use" members?

    _____

60. Is it (True or False) that environmental backlash has reached the political agenda?

Private Land Trusts

61. The purpose of the **private land trust** is to

    a. obtain development rights on private land. (True or False)
    b. purchase land to protect it from development. (True or False)
    c. obtain land at the expense of the landowners. (True or False)
    d. use the power of eminent domain to obtain land. (True or False)

62. Identify a private land trust. _____

63. How many acres are protected by land trusts nationally? _____ acres

Directions: Match each term with the list of definitions and examples given in the tables below. Term definitions and examples might be used more than once.

Term	Definition	Example (s)	Term	Definition	Example (s)
carrying capacity			natural resources		
commons			natural services		
conservation			preservation		
environmental backlash			private land trust		
environmental accounting			renewable resource		
extractive reserves			restoration ecology		
fishery			tragedy of the commons		
maximum sustained yield					

VOCABULARY DEFINITIONS

a. land that is protected for native people

b. a limited marine area or a group of fish species being exploited

c. services provided by ecosystems

d. natural ecosystems and the biota in them

e. including depreciation of natural resources

f. capacity to replenish itself through reproduction

g. manage or regulate use so capacity of species to renew itself is not exceeded

h. ensure ecosystem continuity regardless of potential utility

i. resource owned by many people in common or no one

j. exploitation of the commons leading to a loss of the resource

k. highest possible rate of use that is matched with rate of replacement

l. maximum population the ecosystem can support on a sustainable basis

m. repair damaged ecosystems to normal functions an representative flora and fauna

n. people and organizations that oppose environmental action

o. a nonprofit organization that will accept outright gifts of land or easements

VOCABULARY EXAMPLES	
a. 7.5 million acres of rainforest	f. Fig. 19-11
b. George's Bank, MA	g. Kissimmee River
c. Fig. 19-8	h. National Wetlands Coalition
d. Table 19-3	i. Nature Conservancy
e. Muriquis Monkey	

SELF-TEST

Match the following descriptions with one of nature's services.

Descriptions   Services

1. preventing eutrophication
2. providing plant cover and litter
3. providing natural predators
4. nutrient storage

5. Which of the following actions or statements is not characteristic of the tragedy of the commons?

   a. Whoever harvests the most gains an economic advantage over those who harvest less.
   b. "If I don't get it, someone else will."
   c. Whenever two or more independent groups are engaged in exploitation of a resource.
   d. Reduced harvests leading to reduced attempts to access the resource.

6. Is it (Conservation, Preservation) that implies complete denial of any use of the natural resource?

7. Maximum sustained yield occurs at the optimal population level. What is the **optimum population level**?

8. Is it (Ecological or Economic) reasons why forests are cleared?

   Indicate whether the following are long [L] or short-term [S] consequences of clearing forests.

9. [   ] loss of biodiversity
10. [   ] soil erosion
11. [   ] loss of nutrient recycling
12. [   ] profits from housing developments

13. Which of the following conditions prevent the destruction of tropical rainforests?

   a. Accumulation of huge national debts in Third World countries.
   b. Colonization of rainforest habitats by peasant farmers.
   c. Development of extractive reserves.
   d. Development of pasture lands.

14. Explain why the ocean fishery was considered a commons.

   _____

   _____

15. In the last decade, consumptive use of whales has (Increased, Decreased) and nonconsumptive use has (Increased, Decreased).

16. Identify one nonconsumptive use of whales. _____

17. Explain how a federal redefinition of "wetland" decreased the amount of protected acreage.

   _____

   _____

Match the organizations with the ecosystems under their care.

18. [   ] land trusts   a. National Marine Fisheries Service
19. [   ] National Wildlife Refuges   b. Nature Conservancy
20. [   ] ocean fishery   c. U.S. Fish and Wildlife Service

List four patterns of human uses of natural resources.

21. _____   22. _____

23. _____   24. _____

CHAPTER 20

CONVERTING TRASH TO RESOURCES

When did you last take out the garbage? How often do you perform this task? It is probably a task that you perform frequently, but with little thought about what happens to your garbage. Few people spend time thinking deep thoughts about garbage! This chapter is going to focus your attention on what does happen to your garbage in terms of how it is disposed and, more importantly, what are the ecological impacts of a society that has a "throw-away" mentality. Imagine your personal reaction if the next designated landfill was in your front yard. You would probably object vehemently to such a ridiculous notion. But can you appreciate, in turn, that the landfill you presently use is in nature's front yard?

Make a mental list of all the newspapers, appliances, chemicals, cans and bottles, packaging materials, clothing, and food that you have discarded over the past week. Multiply your volume of refuse by each household on your block, by each household in your town, and finally by each household in the United States. The total is an immense tonnage of wastes on a one-way trip! Where does it all go? The brutal truth is that we are on the brink of a disaster because we are ever increasing our production of trash and are rapidly running out of places to put it. What would you do with your pile of trash if the garbage truck did not show up? In this chapter, we will examine the dimensions of the refuse crisis and some possible solutions to the crisis.

STUDY QUESTIONS

The Solid Waste Problem
Disposing of Municipal Solid Waste

1. List four factors that have contributed to ever increasing amounts of MSW in the U.S.

    a. _____   b. _____

    c. _____   d. _____

2. The three largest components of municipal solid waste are _____, _____, and _____.

3. The proportions of the components of refuse change in terms of

    a. the generator b. the neighborhood c. time of year d. all of these factors

4. Who assumes responsibility for collecting and disposing of municipal solid wastes?

    _____

5. The early form of municipal solid waste disposal was open _____ which were later converted to _____.

6.   What proportions of the MSW in the U.S. are:

     a. dumped in landfills? _____   b. recycled? _____

     c. combusted? _____

Landfills
Problems of Landfills

7.   Indicate whether the following examples are related to the landfill problems of [a] leachate generation, [b] methane production, [c] incomplete decomposition, or [d] settling.

     [   ] explosions
     [   ] groundwater contamination
     [   ] development of shallow depressions
     [   ] "witches brew"
     [   ] biogas
     [   ] 30-year old newspapers

Improving Landfills

8.   Indicate whether the following landfill regulations address [a] leachate generation, [b] methane production, or [c] both a and b

     [   ] Siting landfills on high ground well above the water table.
     [   ] Contoured floors to drain water into tiles.
     [   ] Covering the floor of the landfill with 12 inches of impervious clay or a plastic liner.
     [   ] A gravel layer that surrounds the entire fill.
     [   ] Shaping the refuse pile into a pyramid.
     [   ] Monitoring wells.

9.   One of the most formidable problems facing city governments is _____.

10. What is the implication of *NIMBY*?

     _____

11. List two undesirable consequences of **siting**.

     a. _____   b. _____

12. List two positive aspects of the siting problem.

     a. _____   b. _____

Combustion: Waste to Energy

13. Indicate whether the following are advantages [A] or disadvantages [D] of combustion.

     [   ] 90% reduction in waste volume

[ ] generation of electricity
[ ] incinerator effluents
[ ] cost of construction
[ ] competition with recycling efforts
[ ] WTE
[ ] SEMASS
[ ] resource recovery

Costs of Municipal Solid Waste Disposal

14. Increasing costs of disposing of MSW are related to

    a. new design features of landfills. (True or False)
    b. acquiring landfill sites. (True or False)
    c. transportation to landfills. (True or False)
    d. **tipping fees** at landfills. (True or False)

15. Acquiring a landfill site far enough away from suburbia automatically increases
    _____ costs.

16. Indicate whether the following events will [+] or will not [-] happen because of difficulties in
    establishing new landfill locations.

    [ ] Continued use of old landfills with inadequate safeguards.
    [ ] More old landfills closed than can be replaced with new landfills.
    [ ] A landscape covered with pyramids.
    [ ] Reduced refuse production by society.

Solutions

17. List three alternatives to landfilling some of the components of refuse.

    a. _____   b. _____

    c. _____

Reducing Waste Volume

18. The two largest contributors to waste volume are _____ products and excessive
    _____.

19. _____ items in their existing capacity is the most efficient form of recycling.

20. Identify two methods of reusing items in their existing capacity.

    a. _____   b. _____

Returnable Versus Nonreturnable Bottles

21.   Nonreturnable containers constitute what percent of:

   a. solid waste stream in the U.S.? _____
   b. nonburnable portion of MSW? _____
   c. roadside litter? _____

22.   Larger profits are generated from nonreturnable containers through

   a. reduced transportation costs. (True or False)
   b. indefinite bottle production. (True or False)
   c. eliminating the returnable bottle competition. (True or False)
   d. opposing bottle laws. (True or False)

23.   Bottle laws encourage the use of (Returnable, Nonreturnable) containers.

24.   Bottle laws will (Decrease, Increase) jobs and (Decrease, Increase) litter.

25.   How many states have bottle laws? _____

26.   Who are the greatest opponents to bottle bills? _____

Other Measures

27.   List two other measures to reduce the MSW stream.

   a. _____   b. _____

The Recycling Solution

28.   What percent of the MSW is recyclable? _____ %

29.   Match the below types of refuse with the type of product that reprocessing will produce.

Refuse Type	Product
[ ] paper	a. compost
[ ] glass	b. refabrication without ore extraction
[ ] plastic	c. synthetic lumber
[ ] metals	d. substitute for gravel and sand
[ ] food and yard wastes	e. cellulose insulation
[ ] textiles	f. strengthening agent in recycled paper

Municipal Recycling

30.   List the six characteristics of an effective municipal recycling program.

   a. _____

b. _____

c. _____

d. _____

e. _____

f. _____

Paper Recycling

31. What proportion of the MSW stream is newspaper? _____ %

32. Is it (True or False) that more paper is being recycled than landfilled?

33. A ton of recycled newspaper = _____ trees.

34. The market price for a ton of used newspaper is $_____.

Plastic Recycling

35. Is it (True or False) that all plastic containers are recyclable?

36. List the products made from recycled:

PETE: _____

HDPE: _____

Composting

37. Composting produces a residue of decomposition called _____.

38. Is it (True or False) that there are business opportunities in composting?

Public Policy and Waste Management
The Regulatory Perspective

39. Match the federal legislation with the role each plays in addressing MSW management.
   a. Solid Waste Disposal Act of 1965,   b. Resource Recovery Act of 1970,
   c. Resource Recovery Act of 1976,   d. Superfund Act of 1980,
   e. Hazardous and Solid Waste Amendments of 1984

   [ ] gave EPA greater responsibility to set solid-waste criteria for all hazardous-waste facilities
   [ ] gave jurisdiction over solid waste to Bureau of Solid Waste Management
   [ ] encouraged states to develop some kind of waste management program
   [ ] gave EPA power to close local dumps and set regulations for landfills
   [ ] addressed abandoned hazardous waste sites

Integrated Waste Management

40.   Integrated waste management uses (One, Several) recycling methods.

41.   List three recommendations for MSW management that offer the greatest opportunity to promote sustainability.

a. _____   b. _____

c. _____

VOCABULARY DRILL

Directions: Match each term with the list of definitions and examples given in the tables below. Term definitions and examples might be used more than once.

Term	Definition	Example (s)	Term	Definition	Example (s)
landfill			secondary recycling		
MSW			SEMASS		
primary recycling			waste-to-energy		
resource recovery					

VOCABULARY DEFINITIONS

a. surface repositories for MSW

b. conversion of MSW to electricity

c. separating and recovering materials from MSW

d. modern resource recovery facility

e. original waste material made back into the same material

f. total of all the materials thrown away from homes and commercial establishments

g. waste materials made into different products

VOCABULARY EXAMPLES

a. Fig. 20-4	d. trash, refuse, garbage
b. Fig. 20-8	e. newspapers to cardboard
c. newspapers to newsprint	

SELF-TEST

1. Which of the following has not been a cause of increasing volumes of MSW?

   a. changing lifestyles b. disposable materials c. recycling d. increased gross national product

2. The three largest categories of municipal solid wastes are

   a. paper, cans, and bottles.
   b. paper, yard waste, metal
   c. plastics, paper, and metals.
   d. food, paper, and plastics.

3. The proportions of the components of refuse change in terms of

   a. the generator   b. the neighborhood   c. time of year   d. all of these

4. Most municipal solid waste in the United States is presently disposed of

   a. by burning it in a closed incinerator.
   b. in landfills.
   c. by barging it to sea and dumping it.
   d. in open-burning dumps.

5. The potential of groundwater contamination from landfills is primarily due to

   a. leachate   b. methane gas c. settling   d. explosions

6. Newspaper into newsprint is an example of (Primary, Secondary) recycling?

7. Increasing costs of landfilling are related to

   a. new design features b. acquiring new sites c. transportation d. all of these

8. Even though establishing new landfill locations is extremely difficult, which of the following events will not occur in spite of this difficulty?

   a. Continued use of old landfills with inadequate safeguards.
   b. More old landfills closed than can be replaced with new landfills.
   c. Reduced refuse production by society.
   d. A landscape covered with pyramids.

9. Converting PETE into carpet fiber is an example of (Primary, Secondary) recycling?

10. In term of recycling potential, paper can be

    a. repulped and made into paper and paper products.
    b. manufactured into cellulose insulation.
    c. composted to make a nutrient-rich humus.
    d. all of the above

11. Which of the below examples of recycling laws places the burden mainly on the consumer rather than the government?

    a. mandatory recycling
    b. advance disposal fees
    c. bans on the disposal of certain items
    d. mandating government purchase of recycled materials

12. Burning raw refuse does not eliminate which of the following problems of recycling?

    a. sorting   b. hidden costs c. reprocessing d. marketing

13. Which of the following is not a potential value derived from incinerated refuse?

    a. metal salvage   b. returnable bottles   c. extended use of existing landfills
    d. fill dirt for construction sites

14. Which of the following is not a potential value of "bottle bills"?

    a. Reduced transportation costs.
    b. Indefinite bottle production.
    c. Increased jobs.
    d. Decreased litter.

15. Which of the following measures reduce the amount of material going into trash?

    a. buying less   b. decreased obsolescence   c. yard sales   d. all of these

CHAPTER 21

ENERGY RESOURCES: THE RISE (AND FALL?) OF FOSSIL FUELS

In order to comprehend the importance of Chapter 21 content, develop an inventory of all your direct and indirect reliance on energy just to maintain your quality of life for one day. For example, consider the energy required to maintain a comfortable environment, produce and prepare food, provide transportation, and to produce the diversity of goods and services we all enjoy. Do you know the source of our energy - what we now consider our energy reserves? It is not the sun or any other sustainable supply. We have near total reliance on fossil fuels as our reserve of potential energy. As you know from previous chapters, fossil fuels are nonrenewable and nonrecyclable. In short, we rely almost totally on an energy reserve that is bound to become exhausted, perhaps in our lifetime. To compound the problem, our energy demands increase proportionally to population increase and consumer demand for new and different products and services. We are clearly on a path to nonsustainable energy development.

What are the expected outcomes of nonsustainable energy development? Try to imagine how dramatically your quality of life would change if our energy supplies were interrupted for 48 hours. For most, this would mean no electricity, no heating or air conditioning, no gasoline, no hot baths, no lights, no elevator service, no reason to go to work, and a multitude of other dramatic and negative results. Our energy reserves are one-way trips which means once they are used they are gone forever. One can therefore predict that a quality of life that is dependent on nonrenewable energy reserves has a life span that is directly correlated to the amount of the reserve. When the reserve runs out, the quality of life ends. The purpose of this chapter is to provide a clear picture of the nonsustainability of current energy resources.

STUDY QUESTIONS

Energy Sources and Uses
Harnessing Energy Sources: An Overview

1. Throughout human history, the major energy source was _____ labor.

2. The primary limiting factor to machinery designs in the early 1700s was a _____ source.

3. The _____ engine launched the Industrial Revolution in the late 1700s.

4. The first major fuel for steam engines was (a. coal, b. wood, c. gasoline, d. electricity).

5. By the 1800s the major fuel for steam engines was (a. coal, b. wood, c. gasoline, d. electricity).

6. By 1920, coal provided _____ percent of all energy used in the United States.

7. Three 1800 technologies that provided an alternative to coal were the _____ engine, _____ drilling technology, and the ability to refine _____ oil.

8.	Crude oil provides _____ percent of the total U.S. energy demand.

9.	Coal provides _____ percent of the total U.S. energy demand.

10.	Indicate whether the following attributes are [A] advantages, or [D] disadvantages of oil-based fuels when compared to coal.

	[ ] portability [ ] pollution [ ] energy content [ ] known reserves

Electrical Power Production

11.	Electricity is a (Primary, Secondary) energy source.

12.	One **primary energy source** for generating electrical power is (a. coal b. wood c. oil-based fuels).

13.	Indicate whether the following are advantages [A] or disadvantages [D] of electrical power.

	[ ] pollution from secondary energy source
	[ ] pollution from primary energy sources
	[ ] habitat alterations
	[ ] environmental effects of mining and processing
	[ ] efficiency of secondary energy source

14.	Thermal production of electricity is only _____ % efficient.

15.	_____ % of the primary energy is lost in the form of _____ causing _____ pollution in aquatic ecosystems.

16.	Explain how the below forms of generating electricity represent "transferral of pollution from one place to another."

	Coal-burning power plants: _____

	Hydroelectric power plants: _____

Connecting Sources to Uses

17.	Identify the four major categories of energy use (see Figure 21-8).

	a. _____	b. _____

	c. _____	d. _____

18.	What percent of the total energy use is devoted to:

	a. transportation. _____ %
	b. industry. _____ %
	c. residential. _____ %

d. electrical power generation. _____ %

19.   Match the dominant primary energy sources with list of secondary energy uses.

        a. oil-based fuels   b. natural gas   c. coal, d. nuclear power and other sources
        e. more than one of these energy source is used for the stated purpose

        [ ] transportation
        [ ] industrial processes
        [ ] space heating and cooling
        [ ] generation of electrical power

20.   The energy source that supplies over 40 percent of our total energy and virtually 100 percent of
        our transportation is (a. oil-based fuels, b. natural gas, c. coal, d. nuclear power and other
        sources).

21.   All nuclear power and virtually all coal is used to generate _____ power.

Declining Reserves of Crude Oil
How Fossil Fuels Are Formed

22.   Three examples of fossil fuels are:

        a. _____   b. _____   c. _____

23.   Fossil fuels are accumulations of organic matter from rapid _____ activity
        that occurred at various times in the Earth's history.

24.   Indicate whether the following statements are true [+] or false [-] concerning fossil fuels.

        [ ] Supplies are limited.
        [ ] The significant accumulations of organic matter that produced them can happen today.
        [ ] We are using them far faster than they can be produced.

Crude Oil Reserves vs. Production

25.   Amounts of crude oil that are estimated to exist in the earth are called **estimated**

        _____.

26.   Oil fields that have been located and measured are called _____ **reserves**.

27.   The maximum annual production from a given oil field is limited to about _____ percent
        of the (Proven, Remaining) reserves.

        Assume that a field has 100 million barrels of recoverable reserves. Given a maximum
        production of 10 percent, how much can be withdrawn in

28.   Year 1: _____ barrels leaving _____ barrels

29.   Year 2: _____ barrels leaving _____ barrels

30.   Is it (True or False) that production from a given field may continue indefinitely?

31.   Is it (True or False) that as maximum production goes down, the price per barrel will increase?

Declining U.S. Reserves and Increasing Importation
The Oil Crisis of the 1970s

32.   Indicate whether the following events have increased [I] or decreased [D] in the United States since the 1970s.

        [ ] Consumption of fuels derived from oil.
        [ ] Discoveries of new oil in the United States.
        [ ] Production of oil in the United States.
        [ ] The gap between production and consumption.
        [ ] United States dependence on foreign oil.

33.   **OPEC** stands for

        _____

34.   The OPEC nations created an oil "crisis" in the 1970s by (Increasing, Suspending) oil exports.

35.   Through its ability to create world oil shortages, OPEC (Increased, Decreased) the price of crude oil.

36.   Indicate whether increasing the price of crude oil caused the following events to increase [I] or decrease [D] in the United States.

        [ ] The rate of exploratory drilling and discovery of new oil.
        [ ] Renewed production from old oil fields.
        [ ] Efforts toward fuel conservation.
        [ ] Consumption.
        [ ] Development of alternative energy sources.
        [ ] Dependence on foreign oil.

37.   Is it (True, False) that the results of the oil "crisis" were not shortages but an oil glut?

38.   As a result of oil surpluses, the price of a barrel of crude oil (Increased, Decreased).

39.   As a result of oil surpluses, the price of a gallon of gas at the gas station

        a. increased   b. decreased   c. stayed at the same level as during the crisis.

Victims of Our Success

40.   Indicate whether the collapse in oil prices has caused the following events to increase [I] or decrease [D] in the United States.

        [ ] The rate of exploratory drilling and discovery of new oil.
        [ ] Renewed production from old oil fields.
        [ ] Efforts toward fuel conservation.

[ ] Consumption.
[ ] Development of alternative energy sources.
[ ] Dependence on foreign oil.

41.   U.S. dependence on foreign oil in 1994 has grown to _____ percent.

42.   Is it (True or False) that OPEC oil reserves will last forever.

Problems of Growing U.S. Dependency on Foreign Oil

43.   List three problems resulting from U.S. Dependency on foreign oil.

        a. _____   b. _____

        c. _____

Costs of Purchase

44.   The cost for foreign crude oil represents about _____ % of the U.S. balance trade deficit.

45.   Compare the economic impact of using:

        oil produced at home. _____

        foreign oil. _____

Risk of Supply Disruptions

46.   The Persian Gulf War was fought primarily to:

        a. free the people of Kuwait.
        b. protect Kuwait oil fields from Saddamm Hussein
        c. drive oil prices up.
        d. force OPEC nations to come to terms on oil prices.

47.   Did military intervention (Increase, Decrease) the price/barrel of oil?

Resource Limitations

48.   The last major oil discovery in the U.S. was in _____ in 1968.

49.   The proven oil reserves amounted to another _____ years supply in 1970 compared to a _____ year supply in 1995.

Alternative Fossil Fuels

50.   Unlike oil, the U.S. is rich in supplies other fossil fuels including:

        a. _____   b. _____   c. _____

51. Identify the fossil fuel (a. natural gas, b. coal, c. **synfuels**) that ranks the highest in each of the following categories.

  [ ] air pollution
  [ ] cost of extraction
  [ ] available reserves
  [ ] greenhouse effect
  [ ] retrofitting to existing transportation technologies
  [ ] habitat alteration

Global Warming - A Limiting Factor

52. In the United States, just _____ % of the world's population produces _____ % of the global CO_2 emissions.

53. Which fossil fuel produces the least CO_2 emissions? _____

Sustainable Energy Options

54. Identify two options to combat future oil shortages.

  a. _____   b. _____

Conservation

55. A major preparation for future oil shortages is to (Increase, Decrease) oil consumption.

56. One way oil consumption can be decreased is through _____.

57. Is it (True or False) that between 1970 and 1980 the U.S. economy expanded without a corresponding rise in energy consumption?

58. At least _____ percent of the conservation reserve remains to be tapped.

59. List four ways to increase the conservation reserve.

  a. _____

  b. _____

  c. _____

  d. _____

Development of Non Fossil Fuel Energy Sources

60. Identify two major pathways for developing alternative energy sources.

  a. _____

b. _____

NO VOCABULARY DRILL FOR THIS CHAPTER!

SELF-TEST

1.   Rank (1 to 5) the following fuel types in their order of historical use by human society.

[ ] coal   [ ] wood   [ ] electricity   [ ] gasoline   [ ] natural gas

Classifying each energy type in terms of (a. primary b. secondary) source and (a. renewable b. nonrenewable).

	Energy Types	Primary or Secondary	Renewable or Nonrenewable
2.	wood	[ ]	[ ]
3.	coal	[ ]	[ ]
4.	oil-based	[ ]	[ ]
5.	natural gas	[ ]	[ ]
6.	electricity	[ ]	[ ]
7.	nuclear	[ ]	[ ]
8.	synfuels	[ ]	[ ]
9.	solar	[ ]	[ ]

Classify each energy type in terms of primary end uses including (a. transportation b. industry c. residential d. electric power generation e. not used)

10.   [   ] wood
11.   [   ] coal
12.   [   ] oil-based
13.   [   ] natural gas
14.   [   ] electricity
15.   [   ] nuclear
16.   [   ] synfuels
17.   [   ] solar

Indicate which energy types have **all three advantages** of low cost of extraction, low pollution, and high known reserves. (Check all that apply)

18.   [   ] wood   22. [   ] electricity
19.   [   ] coal   23. [   ] nuclear
20.   [   ] oil-based   24. [   ] synfuels
21.   [   ] natural gas   25. [   ] solar

26.   Which of the above energy types is a primary source, renewable, not used, and has all three advantages listed? _____

27.   Explain why this energy source is not used.

_____

_____

Define "fossil" fuel and give two examples.

28.   "Fossil" fuel:

_____

Two examples are: 29. _____ and 30. _____

Explain why fossil fuels are "nonrenewable" and "nonrecyclable."

31.   Nonrenewable: _____

32.   Nonrecyclable: _____

33.   Explain why electric heat systems represent a **double** energy waste:

_____

Indicate whether the following actions will increase [+], decrease [-], or have no impact [0] on future supplies of oil-based fuels.

34.   [   ] searching for new reserves
35.   [   ] taking OPEC oil by force
36.   [   ] increasing the price of gasoline at the pump
37.   [   ] increased use of mass transit systems
38.   [   ] continuing consumption at present rates
39.   [   ] car pooling
40.   [   ] increasing engine fuel efficiency
41.   [   ] increasing the beginning driving age to 18

42.   Identify an oil field that has a potential production of 6 million barrels per day, is three times the size of the Alaskan field, is inexhaustible, and its exploitation will not adversely effect the environment.

This oil field is called the _____ reserve.

CHAPTER 22

NUCLEAR POWER: PROMISE AND PROBLEMS

You have probably now come to realize that we cannot rely solely on oil-based fuels to maintain our quality of life into the future. In fact, it would be both foolhardy and disastrous to perpetuate such a narrow energy policy. Now that you have been convinced of the perils facing an oil-based society, it is fair to ask, "What are the alternatives?" It would be wonderful to give society a long shopping list of energy alternatives, but given our present technology, this is not possible. We have several alternative technologies that are viable only because (1) they are linked to energy producing resources that are more abundant than oil and (2) they can only be used to produce electricity. These alternatives are of little value in the transportation industry unless automobiles can be designed to run on different types of energy reserves, (e.g., electricity). There are other limitations to these alternatives.

Nuclear power was, at one time, considered the panacea for all our energy needs. This unguarded optimism was the result of no long-term experimentation on the promise and problems of nuclear power. Today, we are well aware of the extreme danger that can be reaped upon whole populations and ecosystems given a cavalier approach to the safety requirements in the use of nuclear power (e.g., Chernobyl and Three Mile Island). In fact, people now have almost a paranoid fear of nuclear reactors. This is unfortunate because nuclear power does have great potential as an alternative means of generating electricity. This potential will be realized, even accepted, if the safety technology becomes as well developed as the power generating technology. The purpose of this chapter is to investigate the potential of nuclear power.

STUDY QUESTIONS

Nuclear Power: Dream or Delusion?

1. In the 1960s and early 1970s, the perception of nuclear power plants was one of unguarded (Optimism, Pessimism).

2. Since 1975, the perception of nuclear power plants was one of (Optimism, Pessimism).

3. What factor has the greatest impact on future developments of nuclear power plants?

    a. cost b. government financing c. public opinion d. foreign investments

4. What country produces 73% of its electricity from nuclear power (see Fig. 22-4)?

    a. United States   b. Japan   c. France   d. Russian Republics

5. Is it (True or False) that impending oil shortages are real?

How Nuclear Power Works

6.   State the objective of nuclear power technology.

        _____

From Mass to Energy

7.   Which of the following is a definition of [a] **fission** or [b] **fusion**?

        [   ] A large atom of one element is split into two smaller atoms of different elements.
        [   ] Two small atoms combine to form a larger atom of a different element.

8.   In fission and fusion, the mass of the products is (More, Less) than the mass of the starting
        material.

9.   The lost mass is converted into tremendous amounts of _____.

10.   Controlled fission releases this energy gradually as _____.

The Fuel for Nuclear Power Plants

11.   All nuclear power plants use (Fission, Fusion).

12.   The raw material of nuclear power plants is uranium (235, 238).

13.   Uranium-235 and 238 are both _____ of the element called uranium.

14.   **Isotopes** are defined by the number of (a. protons b. neutrons c. electrons) in an element.

15.   The isotope that will fission is uranium (235, 238).

16.   The "bullet" that starts the uranium-235 fission process is a (a. proton b. neutron c. electron
        d. element)

17.   Three products of a uranium-235 fission are (see Figure 22-5)

        a. _____   b. _____   c. _____

18.   A _____ reaction occurs when the neutrons cause other fissions which release
        more neutrons which cause other fissions.

19.   The majority of all uranium is the (235, 238) isotope.

20.   Enrichment is the process of enhancing the concentration of uranium (235, 238).

21.   A nuclear bomb explosion is the result of a (Controlled, Uncontrolled) fission of (High, Low)
        grade of uranium 235.

The Nuclear Reactor

22.   A nuclear reactor is designed to

a. sustain a continuous chain reaction. (True or False)
b. prevent amplification into a nuclear explosion. (True or False)
c. consist primarily of an array of fuel and control rods. (True or False)
d. make some material intensely hot. (True or False)
e. convert heat into steam. (True or False)
f. convert steam into electricity. (True or False)

23.   Indicate whether the following are functions of [a] the **moderator**, [b] **fuel rods**, or [c] control rods.

[   ] Contain the mixture of uranium isotopes.
[   ] Contain a neutron-absorbing material.
[   ] Starts the chain reaction.
[   ] Controls the chain reaction.
[   ] Stores the heat energy to produce steam.
[   ] Slows down the fission neutrons.

24.   A nuclear reactor is simply an assembly of:

a. _____   b. _____   c. _____

The Nuclear Power Plant

25.   A nuclear power plant is designed to

a. convert heat into steam. (True or False)
b. use steam to drive turbogenerators. (True or False)
c. convert steam into electricity. (True or False)
d. prevent meltdown. (True or False)
e. produce super heated water in a reactor vessel. (True or False)

26.   A loss-of-coolant accident has the potential of causing _____.

27.   The fissioning of a pound of uranium fuel releases energy equivalent to burning _____ tons of coal.

Comparisons of Nuclear Power and Coal Power

28.   Indicate whether the following characteristics pertain to [C] coal-fired or [N] nuclear power plants.

[   ] Requires 3.5 million tons of raw fuel.
[   ] Requires 1.5 tons of raw material.
[   ] Will emit over 10 million tons of carbon dioxide into the atmosphere.
[   ] Will emit no carbon dioxide into the atmosphere.
[   ] Will emit over 400 thousand tons of sulfur dioxide into the atmosphere.

[ ] Will emit no acid forming pollutants.
[ ] Will produce about 100 thousand tons of ash.
[ ] Will produce about 250 tons of radioactive wastes.
[ ] Presents the possibility of catastrophic accidents, e.g., meltdown.

Radioactive Materials and Their Hazards
Radioactive Emissions

29. The fission products of unstable isotopes are called _____.

30. **Radioisotopes** gain stability by ejecting _____ particles or high-energy _____.

31. Subatomic particles and high-energy radiation is referred to as _____ **emissions.**

32. The indirect and direct products of fission are the _____ **wastes** of nuclear power.

33. Is it (True or False) that other materials in and around the reactor may become unstable isotopes?

34. Unstable isotopes are (Direct, Indirect) products of fission.

Biological Effects

35. List five biological effects of radioactive emissions.

    a. _____

    b. _____

    c. _____

    d. _____

    e. _____

Sources of Radiation

36. Radioactive emissions from nuclear power plants that are operating normally are generally (Higher, Lower) than background radiation.

37. List two sources of **background radiation.**

    a. _____   b. _____

38. Public exposure to radiation from normal operations of a power plant is < _____ % of natural background.

39.	List the two real public concerns regarding radiation emissions from nuclear plants.

	a. _____	b. _____

Radioactive Wastes

40.	The process whereby unstable isotopes eject particles and radiation is known as _____ **decay**.

41.	The main variable that determines the rate of radioactive decay of a given isotope is its _____ **life**.

42.	Half-lives vary from a fraction of a _____ to _____ of years.

43.	Which of the following isotopes has the shortest [S] and which has the longest [L] half-life? (see Table 22-2)

	[ ] Iodine-131,		[ ] Cesium-137,		[ ] Plutonium-239

44.	Which of the above isotopes would require the

	shortest containment time? _____

	longest containment time? _____

45.	Generally, (1, 10, 100) half-lives are required to reduce radiation levels to insignificant levels.

46.	To be safe, plutonium-239 (half-life = 24,000 years) would require _____ thousand years of containment.

Disposal of Radioactive Wastes

47.	Radioactive waste disposal falls into the two categories of _____ term and _____ term containment.

48.	The major problems of radioactive waste disposal are related to

	a. finding long-term containment sites. (True or False)
	b. transport of highly toxic radioactive waste across the United States. (True or False)
	c. the lack of any resolution to the radioactive waste problem. (True or False)

49.	The world's commercial reactors have accumulated _____ thousand tons of nuclear wastes.

50.	List two types of nuclear wastes that present serious disposal problems.

	a. _____	b. _____

Military Radioactive Wastes

51.   Of what significance are the following two locations concerning radioactive waste disposal.

    a. Savannah River: _____

    b. Lake Karachay: _____

52.   What is the estimated cost of cleaning-up and stabilizing the 20 nuclear weapons- production sites in the U.S.? $ _____ billion to $ _____ trillion.

High-level Nuclear Waste Disposal

53.   What is the significance of the acronym NIMBY in the radioactive disposal debate?

    _____

54.   What area of the United States might become the nation's nuclear dump? _____

The Potential for Accidents

55.   The catastrophic ramifications of the nuclear accident at Chernobyl were intensified due to the lack of a _____ building.

56.   The near catastrophic nuclear accident at Three Mile Island was due to lack of _____ in the reactor core.

57.   In both the Chernobyl and Three Mile Island cases, what was the fundamental cause of both accidents?

    _____

58.   Has public confidence in nuclear engineering and engineers (Increased, Decreased) since Chernobyl and Three Mile Island.

59.   Are nonsmokers (Equally, Less, More) likely to suffer the effects of radiation as smokers? Answer is not in the book, but think about it!

Safety and Nuclear Power

60.   Nuclear proponents now claim that nuclear reactors are _____ times safer.

61.   Nuclear proponents are attempting to increase public confidence in nuclear power through

    a. **active safety** designs. (True or False)
    b. **passive safety** designs. (True or False)
    c. advanced light water reactors. (True or False)
    d. professing the very low probabilities of nuclear accidents. (True or False)

Economic Problems With Nuclear Reactors

62.   Utilities are turning away from nuclear power primarily because of

a. adverse public opinion   b. government constraints
c. increased costs   d. concern for the environment

63.   The cost of building a nuclear power plant has escalated due to

a. new safety standards. (True or False)
b. construction delays. (True or False)
c. shorter-than-expected lifetime of nuclear power plants. (True or False)
d. withdrawl of government subsidies. (True or False)
e. the demand for electricity being met by coal-fired power plants. (True, False)

64.   List two factors that have contributed to shorter-than-expected lifetimes of nuclear power plants.

a. _____   b. _____

More Advanced Reactors

65.   Identify two alternative nuclear power technologies.

a. _____   b. _____

66.   Indicate whether the following are characteristics of breeder [BR], fusion [FU], or both [BO]
types of reactors.

[   ] Creates more fuel than it consumes.
[   ] The raw material is uranium-238.
[   ] Splits atoms.
[   ] Produces radioactive wastes.
[   ] Produces plutonium-239 as radioactive waste.
[   ] Fuses atoms.
[   ] Releases energy.
[   ] The raw material is deuterium and tritium.
[   ] The source of unprecedented thermal pollution.

The Future of Nuclear Power

67.   Public opposition to nuclear energy is based on:

a. _____

b. _____

c. _____

d. _____

VOCABULARY DRILL

Directions: Match each term with the list of definitions and examples given in the tables below. Term definitions and examples might be used more than once.

Term	Definition	Example (s)	Term	Definition	Example (s)
active safety			half-life		
background radiation			indirect products		
breeder reactor			isotopes		
corrosion			LOCA		
direct products			LWRs		
embrittlement			mass number		
enrichment			passive safety		
fission products			radioactive emissions		
fission			radioactive decay		
fuel rods			radioactive wastes		
fusion			radioisotopes		
fusion reactor			rem		

VOCABULARY DEFINITIONS

a. a large atom of one element is split to produce two different smaller elements

b. two small atoms combine to form a larger atom of a different element

c. different forms of the same element

d. particles or rays emitted during fission

e. the number of neutrons and protons in the nucleus of the atom

f. process of producing a material containing a higher concentration of a desired element

g. rods full of uranium pellets

h. loss-of-coolant accident

i. unstable isotopes of the elements resulting from the fission process

j. materials that become radioactive by absorbing neutrons from the fission process

k. ability for radioactive emission to do biological damage

l. radiation that occurs naturally in the earth's crust

VOCABULARY DEFINITIONS
m. when unstable isotopes eject particles and radiation and become stable and nonradioactive
n. time for half the amount of a radioactive isotope to decay
o. safety that relies on operator-controlled actions
p. safety that relies on engineering devices and structure
q. effect on materials from constant neutron bombardment
r. effect on plumbing from corrosive chemicals
s. reactors that produce fuel in the process of consuming fuel
t. reactors that use H sources as fuel elements
u. reactors that use very pure water as the reactor moderator

VOCABULARY EXAMPLES	
a. cracked metal or pipes	i. Table 22-2
b. alpha and beta particles, neutrons	j. Superphenix
c. Fig. 22-8	k. Fig. 22-7
d. Fig. 22-5a	l. Fig. 22-14
e. Fig. 22-5b	m. humans
f. uranium-235	n. Table 22-1
g. Fig. 22-10	o. Three Mile Island
h. radon gas	

SELF-TEST

1. Which of the below factors will have the greatest impact on determining future developments of nuclear power?

    a. cost b. government financing c. public opinion d. foreign investments

2. Which of the following best describes nuclear fission?

    a. A large atom of one element is split into 2 smaller and different atoms with the resulting release of free neutrons and energy.
    b. Two small atoms are melted together to form a larger atoms with some mass being converted to energy.
    c. Energy is used to split a large atom into 2 smaller atoms with resulting release of energy.
    d. All of the above are proper explanations of fission.

3.   The fuel for present nuclear power reactors is

a. uranium-235 b. uranium-238 c. plutonium-239 d. thorium

4.   The "bullet" that starts the uranium-235 fission process is

a. a proton b. a neutron c. an electron d. an atom

5.   What is the process called that results in the splitting of one uranium atom that releases free neutrons causing other uranium atoms to split which in turn releases more neutrons causing other uranium atoms to split?

a. nuclear fusion b. nuclear fission c. chain reaction d. enrichment

6.   The process of enhancing the concentration of uranium-235 is called

a. nuclear fission b. a meltdown c. a chain reaction d. enrichment

7.   Which of the following is not a function of the nuclear reactor?

a. Sustain a continuous chain reaction.
b. Prevent amplification into a nuclear explosion.
c. Convert heat into steam.
d. Make some material intensely hot.

8.   Which of the following is not a function of the fuel rod?

a. Contain a mixture of uranium isotopes.
b. Start a chain reaction.
c. Control a chain reaction.
d. Become intensely hot.

9.   A nuclear power plant is designed to

a. produce super heated water.
b. convert heat into steam.
c. convert steam into electricity.
d. do all of the above.

10.   Which of the following is a disadvantage of nuclear power plants when compared to coal-fired power plants?

a. The amount of raw material required.
b. The type of waste produced.
c. The wattage of electrical power produced.
d. The amount of sulfur dioxide produced.

11. The major disadvantage(s) of using atomic energy to generate electricity is/are that

a. it causes a large amount of thermal pollution.
b. it produces very dangerous waste products.
c. the ultimate goal of putting a nuclear power plant on line is redundant to the goal of coal-fired generating plants.
d. all of the above are disadvantages.

12. Radioactive emissions are not know to cause

a. damage to biological tissues.
b. damage to DNA molecules.
c. a greenhouse effect.
d. increases in background radiation.

13. The major problems of radioactive waste disposal is not related to

a. finding long-term containment facilities.
b. transport of highly toxic radioactive wastes across the United States.
c. a lack of resolution to the radioactive waste problem.
d. background radiation around nuclear power plants.

14. Escalating costs of building a nuclear power plant are not related to

a. environmental concerns.
b. new safety standards.
c. construction delays.
d. embrittlement.

15. The lack of public support for nuclear technology is based on

a. distrust of the technology and those who support it.
b. resulting environmental pollution problems.
c. their inability to understand the technology.
d. all of the above

CHAPTER 23

SOLAR AND OTHER RENEWABLE ENERGY SOURCES

Here is the ultimate trivia question. Can you identify an energy source that is abundant, everlasting, and nonpolluting? You could ask this question of someone while standing outside on a bright sunny day and they would not be able to identify the sun as the correct answer. It is indeed ironic that, given the tremendous energy needs of our society both now and in the future, we have failed to utilize the one energy source that meets the very criteria we seek. The main barrier to full-scale utilization of solar energy is economic. No one knows how to make a profit from selling sunlight! Does this surprise you? You may have thought that the economic barrier was related to development of the technologies to trap and store solar energy. Your study of Chapter 23 will make you aware of several technologies that are quite efficient, practical, and sophisticated. It is reasonable to assume that stock holders in the oil, coal, or nuclear industries would become quite concerned if the public demanded more government support for solar technology development.

The economic worries of stock holders in the "dirty" power industries will be intensified as greater attention is given to the "clean" alternatives of wind, geothermal, tidal, and wave power. It is amazing that these natural power sources have always been available but drastically underutilized. The main reason they have been ignored is the same as for solar energy. How do you make a profit from the wind and waves? This chapter examines the range of technologies that are available and what is necessary to promote their development and implementation.

STUDY QUESTIONS

Principles of Solar Energy

1. Solar energy is a (Potential, Kinetic) form of energy originating from (Chemical, Thermonuclear) reactions in the sun.

2. Solar energy is a (Renewable, Nonrenewable) form of energy.

3. The four main hurdles to overcome in using solar energy include

   a. _____   b. _____

   c. _____   d. _____

Putting Solar Energy to Work

4. List three direct uses of solar energy.

   a. _____   b. _____

   c. _____

Solar Heating of Water
Solar Space Heating

5.   Indicate whether the below components are required of **active** [A], **passive** [P], or both [B]
        types of solar heating systems. (see Fig. 23-6 to 23-10)

        [   ] **flat-plate solar collector**
        [   ] water pump
        [   ] heat storage facility
        [   ] expansion tank
        [   ] blowers
        [   ] back-up system
        [   ] heat exchanger
        [   ] valves
        [   ] improved insulation
        [   ] appropriate landscaping

6.   Indicate whether the below components are requirements of solar hot water [W], space [S],
        or both [B] types of solar heating systems. (see Fig. 23-6 to 23-10)

        [   ] **flat-plate solar collector**
        [   ] water pump
        [   ] heat storage facility
        [   ] expansion tank
        [   ] blowers
        [   ] back-up system
        [   ] heat exchanger
        [   ] valves
        [   ] improved insulation
        [   ] appropriate landscaping

7.   The most economical means of solar heating is an/a (Active, Passive) system.

8.   A well-designed passive solar home can reduce energy bills by _____ percent with added
        construction costs of _____ to _____ percent.

9.   Is it (True or False) that partial conversion to solar energy will make a large amount of natural
        gas available for fueling vehicles?

10.   About _____ percent of the total U.S. energy budget is used for heating buildings and
        water.

11.   The chief barrier to more wide-spread use of passive solar designs is _____.

Solar Production of Electricity

12.   List two technologies that convert solar energy into electricity.

        a. _____   b. _____

13. A device that converts light directly to electricity is the _____ cell.

14. The photovoltaic or solar cell consists of two layers of

    a. water   b. electrons   c. atoms   d. mylar film.

15. The top layer contains atoms with (Additional, Missing) electrons in the outer orbital.

16. The bottom layer contains atoms with (Additional, Missing) electrons in the outer orbital.

17. The top layer will readily (Lose, Accept) an electron.

18. The bottom layer will readily (Lose, Accept) an electron.

19. The kinetic energy of sunlight forces a movement of electrons from the top to the bottom layer creating an electrical imbalance or _____.

20. Solar cell technology

    a. is becoming more cost-effective with increased use. (True or False)
    b. will not wear out. (True or False)
    c. is already competitive in costs per watt output. (True or False)
    d. has the potential to provide power for anything from a watch to a home. (True or False)

Solar Trough Collectors

21. Solar trough collectors convert _____ percent of incoming sunlight to electric power at a cost of _____ cents per kilowatt-hour.

22. The economic feasibility of the solar trough collector is enhanced by (More, Less) environmental pollution when compared to coal-fired facilities.

23. Power towers use mirrors to convert solar energy to _____ which runs a _____ that creates electricity.

24. Solar ponds consist of a layer of _____ water overlaid with _____ water.

25. Match the following functions of the solar pond [a] brine water and [b] fresh water.

    [   ] absorption of sunlight energy.
    [   ] conversion of sunlight energy into heat.
    [   ] prevents the heat from escaping.

26. The hot brine solution can be

    a. used directly to heat buildings. (True or False)
    b. converted to electrical power. (True or False)

Solar Production of Hydrogen--The Fuel of the Future

27.   Indicate whether the following are benefits [+] or limitations [-] of hydrogen power.

[   ] substitute for natural gas
[   ] substitute for gasoline
[   ] pollution factor
[   ] production technology
[   ] modification of existing transportation technologies
[   ] production costs
[   ] national distribution system

Indirect Solar Energy

28.   List four indirect forms of solar energy.

a. _____   b. _____

c. _____   d. _____

Hydropower

29.   The earliest form of hydropower was the use of (Natural, Artificial) water falls.

30.   The contemporary form of hydropower is the use of (Natural, Artificial) water falls.

31.   Artificial water falls are created by building huge _____.

32.   The artificial water falls of huge dams generate _____ **power**.

33.   Indicate whether the following are benefits [+] or drawbacks [-] of hydroelectric power.

[   ] Level of pollution generated.
[   ] Level of environmental degradation.
[   ] Amount of total energy produced.
[   ] Geographical distribution of energy produced.

34.   Only _____ percent of the nation's rivers remain free-flowing.

Wind Power

35.   The earliest form of harnessing wind power were wind (Mills, Turbines).

36.   The contemporary form of harnessing wind power is wind (Mills, Turbines).

37.   Indicate whether the following are benefits [+] or drawbacks [-] of wind power.

[   ] The size limitations on wind turbines.
[   ] The number of megawatts of electricity produced.
[   ] The level of pollution generated.

[ ] The level of environmental degradation.
[ ] Geographical distribution of energy produced.
[ ] Cost-effectiveness
[ ] Aesthetics

Biomass Energy of Bioconversion

38.   List four examples of biomass energy or bioconversion.

   a. _____   b. _____

   c. _____   d. _____

39.   Indicate whether the following are benefits [+] or drawbacks [-] of biomass energy.

   [ ] Availability of the biomass resource.
   [ ] Access to the biomass resource.
   [ ] Public acceptance and utilization of biomass energy.
   [ ] Past history of human harvests within a maximum sustained yield.

40.   Methane gas is (Direct, Indirect) use of biomass as fuel.

41.   Is it (True or False) that one can get more electrical power from cows than from nuclear power?

42.   Alcohol production is a (Direct, Indirect) use of biomass as fuel.

43.   Alcohol production utilizes (Fermentation, Distillation, Both).

Ocean Thermal Energy Conversion

44.   Indicate whether the following are benefits [+] or drawbacks [-] of **OTEC**.

   [ ] technology
   [ ] cost-effectiveness
   [ ] environmental consequences

Additional Renewable Energy Options

45.   List two additional sustainable energy options.

   a. _____   b. _____

Geothermal Energy

46.   Indicate whether the following are benefits [+] or drawbacks [-] of geothermal energy.

   [ ] If accessible, it is an everlasting energy resource.
   [ ] The consistency in number of megawatts of electricity produced.
   [ ] The level of pollution generated.
   [ ] The level of environmental degradation.

[ ] Geographical distribution of energy produced.
[ ] Cost-effectiveness
[ ] Technology required for extraction.

Tidal Power

47.   Tidal power converts the energy of (Incoming, Outgoing, Both) tides.

48.   Indicate whether the following are benefits [+] or drawbacks [-] of tidal power.

[ ] If accessible, it is an everlasting energy resource.
[ ] The level of pollution generated.
[ ] The level of environmental degradation.
[ ] Geographical distribution of energy produced.
[ ] Cost-effectiveness of the technology.
[ ] Technology required for energy conversion.

Policy for a Sustainable Energy Future

49.   Indicate whether the following are [+] or are not [-] aspects of the existing U.S. energy policy.

[ ] mandates to increase fuel efficiency of cars
[ ] exploitation of public lands for fossil fuel reserves
[ ] subsidies to producers of solar energy or solar energy products
[ ] continued research on and development of renewable energy resources
[ ] continued research on and development of alternative energy technologies
[ ] advocacy of energy conservation
[ ] public education on solar and other renewable energy sources

Term	Definition	Example (s)	Term	Definition	Example (s)
active systems			OTEC		
bioconversion			passive systems		
biomass energy			power gird		
carbon tax			PV cell		
electrolysis			solar trough		
flate-plate collectors			wind farms		
fuel cells			wind turbine		
geothermal energy					

VOCABULARY DEFINITIONS

a. broad box with a glass top and a black bottom with imbedded water tubes
b. moves water or air with pumps or blowers
c. moves water or air with natural convection currents or gravity
d. cell that converts sunlight energy into electricity
e. a network of power lines taking power from generating stations to customers
f. long trough-shaped reflectors tilted toward the sun
g. disassociation of H_2O using electricity
h. devices in which H is recombined with O_2 chemically to produce electric potential
i. wind-driven generator
j. collections of thousands of wind-driven generators
k. deriving energy from present-day photosynthesis
l. concept of using ocean's temperature differences to produce power
m. naturally heated groundwater
n. tax on fuels according to the amount of CO_2 produced during consumption

VOCABULARY EXAMPLES	
a. burning wood	h. Hawaii
b. Fig. 23-11	i. Fig. 23-9
c. Fig. 23-20	j. pumps, valves, blowers
d. Fig. 23-6	k. European countries
e. Fig. 23-25	l. Fig. 23-17
f. overhead electrical lines	m. Fig. 23-23
g. H and O2 bubbles	

SELF-TEST

1.   The biggest problem connected with using solar energy is

   a. there is not enough energy inherent in sunlight to be worth its use.
   b. there is no good means of collecting sunlight energy.
   c. the means of storing it are bulky and expensive.
   d. there are no good ways of converting solar energy into profits.

Match the below terms or concepts with examples of solar and other renewable energy sources.

   Terms and Concepts   Examples

2.   [   ] flat plate collector   a. solar energy
3.   [   ] active solar system   b. additional renewable energy options
4.   [   ] passive solar system   c. indirect solar energy
5.   [   ] photovoltaic cells
6.   [   ] solar trough collector
7.   [   ] power tower
8.   [   ] hydrogen production
9.   [   ] hydropower
10.   [   ] wind turbine
11.   [   ] bioconversion
12.   [   ] OTEC
13.   [   ] geysers
14.   [   ] twice-daily rise and fall

15.   The most cost-effective method of using solar energy for home use is through

   a. passive solar systems.
   b. active solar systems.
   c. photovoltaic solar systems.
   d. selling your excess power back to the utility companies.

16. Which of the following is not a direct means of using solar energy?

   a. Direct conversion to electrical power.
   b. Heating homes and other buildings.
   c. Production of flammable hydrogen gas.
   d. Production of biomass for conversion to flammable liquids.

17. Solar cell technology

   a. cannot become cost effective even with increased use.
   b. will wear out.
   c. is competitive with nuclear power in costs per watt output.
   d. has severe limitations in power output.

18. A major drawback of solar production of hydrogen is

   a. its ability to replace our reliance on natural gas.
   b. its ability to replace gasoline in cars.
   c. its waste products.
   d. the ability to produce free hydrogen atoms.

19. Hydroelectric power cannot provide a large percentage of power in the future because of the

   a. level of thermal pollution generated.
   b. amount of energy produced.
   c. limited geographic distribution of energy produced.
   d. level of air and water pollution generated.

20. A major drawback of wind power is

   a. amount of energy produced.
   b. size limitations of wind turbines.
   c. level of environmental degradation.
   d. cost-effectiveness.

21. The major drawback that geothermal energy, tidal and wave power all have in common is

   a. the number of megawatts of electricity that could be produced is low.
   b. the technology required for energy conversion is not cost-effective.
   c. the level of environmental degradation.
   d. limitations on long-term dependency as an energy source.

22. In view of the nature of U.S. energy problems and associated environmental problems, which of the following areas should be given a higher priority in research and development funding?

   a. fusion power b. solar power c. breeder reactors d. utilization of coal

CHAPTER 24

LIFESTYLE AND SUSTAINABILITY

██

Our lifestyle specifications concerning shelter and adequate space have led to urban sprawl and the consequent environmental degradation. The irony of urban sprawl is that it recreated the very conditions people were attempting to escape. For example, air pollution levels in rural areas are, in many locations, far worse than in the city. The social consequences of urban sprawl can now be documented and have been found to be as insidious as the environmental impacts. Some communities are experiencing a reverse migration where people are returning to the city and reclaiming the traditions, landmarks, and culture that they left. Understanding this phenomenon and other ways people can and are moderating their lifestyles so that they have less environmental impact is the subject of this chapter.

██

STUDY QUESTIONS

From Urban Structure to Urban Sprawl
The Origins of Urban Sprawl

1. List three factors that contributed the most to urban sprawl.

    a. _____   b. _____

    c. _____

2. Urban sprawl is a function of (Time, Distance).

3. Indicate whether the following factors enhanced [+] or had no effect on [-] urban sprawl.

    [   ] Ability to purchase cars.
    [   ] Commuting time.
    [   ] **Highway Trust Fund**
    [   ] Commuting distance.
    [   ] Population growth
    [   ] The lack of planned development.
    [   ] Low rural tax base.
    [   ] The environmental or social consequences.
    [   ] Henry Ford
    [   ] low-interest mortages
    [   ] Highway Revenue Act of 1956

4. How have average commuting distances and time changed since 1960.

    distance _____   time _____

5.   Explain the difference between a **suburb** and **exurb**.

_____

Environmental Impacts of Urban Sprawl

6.   Indicate whether the following consequences of urban sprawl are related to

a. Depletion of Energy Resources
b. Air Pollution, Acid Rain, the Greenhouse Effect, and Stratospheric Ozone Depletion
c. Degradation of Water Resources and Water Pollution
d. Negative Impact on Recreational and Scenic Areas and Wildlife
e. Loss of Agricultural Land

[   ] loss of 2.5 million acres per year
[   ] highways through parks or along stream valleys
[   ] decreases infiltration over massive areas
[   ] increased numbers and densities of cars
[   ] increased commuting miles
[   ] increased roadkills
[   ] three-fold increase in per capita oil consumption

Social Consequences of Exurban Migration

7.   What is the main, negative, societal impact of exurban migration.

a. _____

8.   Exurban migration

a. has profound economic and social impacts. (True or False)
b. segregates the population along economic and/or ethnic lines. (True or False)

9.   Indicate whether the following characteristics do [+] or do not [-] describe the people who are left behind after exurban migration.

[   ] They are the poor, elderly, handicapped, and minority groups.
[   ] They can obtain mortgage loans.
[   ] They reflect a **gentrification** of society.
[   ] They represent the economically depressed subset of society.

Vicious Cycle of Urban Blight

10.   Local governments

a. are responsible for providing societal services. (True or False)
b. operate from budgets derived from property taxes. (True or False)
c. are separate entities in the central city and suburbs. (True or False)

11.   Property values are generally (Higher, Lower) in the central city when compared to suburbs.

12. Property taxes are generally (Higher, Lower) in the central city when compared to suburbs.

13. An **eroding tax base** is characteristic of the (Suburbs, Central City).

14. The quality and quantity of government services is markedly (Better, Worse) in the central city when compared to the suburbs.

15. Exurban migration causes the central city infrastructure to (Deteriorate, Improve).

Economic Exclusion of the Inner City

16. Is it (True or False) that people will never return to the central city from the suburbs?

17. Is it (True or False) that some programs are being implemented to revitalize the urban core?

18. Indicate whether the following events are likely to increase [+] or decrease [-] as a result of urban decay.

    [   ] Industrial development in the central city.
    [   ] Job opportunities in the central city.
    [   ] The tax base of the central city.
    [   ] The unemployment rate of central city residents.
    [   ] Social unrest and crime rates within the central city.
    [   ] The level of education offered to and achieved by central city residents.
    [   ] Accessibility of jobs outside the central city by its residents.

19. Suburban sprawl is a (Sustainable, Nonsustainable) system.

Moving Toward Sustainable Cities
What Makes Cities Livable?

20. The most livable cities around the world have which of the following characteristics? (Check all that apply)

    [   ] reduced outward sprawl
    [   ] reduced automobile traffic
    [   ] improved access by foot or bicycle
    [   ] mass transit
    [   ] high population density
    [   ] heterogeneity of residences and businesses
    [   ] people meet people not cars

21. List three changes used by Portland, Oregon to make this city more livable.

    a. _____

    b. _____

    c. _____

Reining in Urban Sprawl

22.   List the three basic considerations in developing sustainable communities.

    a. _____

    b. _____

    c. _____

23.   Explain the significance of the Intermodal Surface Transportation Efficiency Act of 1991 in promoting suburb sustainability.

    _____

Reversing Urban Blight

24.   Match the program with its mission.

Program   Mission

    [   ] empowerment zones   a. on the job training
    [   ] urban homesteading   b. promote jobs in depressed areas
    [   ] Habitat for Humanity   c. sweat equity

25.   Taxes are usually based on the value of the (Buildings, Land) but in order rehabilitate the inner city, taxes are based on the value of the (Buildings, Land) only.

Making Communities Sustainable

26.   Is it (True or False) that cities can be sustainable?

27.   Which of the following are characteristics of a sustainable community? (Check all that apply)

    [   ] Proximity of people to residences, shops, and workplaces.
    [   ] Need for cars.
    [   ] Use of solar energy.
    [   ] Ability to utilize alternative energy resources.
    [   ] Self-sufficiency in provision of food.
    [   ] Stable population.
    [   ] Cluster development.

Term	Definition	Example (s)	Term	Definition	Example (s)
Directions: Match each term with the list of definitions and examples given in the tables below. Term definitions and examples might be used more than once.					
economic exclusion			gentrification		
empowerment zones			urban decay		
eroding tax base			urban blight		
exurban migration			urban sprawl		
exurbs					

VOCABULARY DEFINITIONS

a. residential areas, shopping malls, and other facilities laced together by multilane highways

b. relocation of people, residences, shopping areas, and workplaces in the city to outlying areas

c. communities farther from cities than suburbs

d. segregation of the population into groups sharing common economic, social, and cultural backgrounds

e. declining tax revenue resulting from declining property values

f. whole process of exurban migration and debilitating economic and social impacts on the inner city

g. lack of access by inner city residents to suburban jobs

h. coordinated efforts by government and business to revitalize designated inner city zones

VOCABULARY EXAMPLES

a. 50% unemployment rates	d. suburbia	f. Fig. 24-6
b. Table 24-1	e. Fig. 24-11	g. Detroit
c. Fig. 24-2		

SELF-TEST

1. The prime factor that initiated suburban growth and has supported urban sprawl is

    a. farmers selling land to developers.   b. widespread ownership of private cars.

    c. decline of the central city.   d. developers buying farms.

2.   Which of the following factors had no effect on the rate of urban sprawl?

        a. ability to purchase cars b. commuting time c. commuting distance d. rural tax base

3.   An environmental consequence of urban sprawl was

        a. depletion of energy resources.   b. air pollution, acid rain, and the greenhouse effect.
        c. loss of agricultural land.   d. all of the above

4.   Which of the following is not a characteristic of the people left behind after exurban migration?

        a. They are the poor, elderly, handicapped, and minority groups.
        b. They have high levels of education and income.
        c. They are predominantly economically depressed minorities and whites.
        d. They reflect the gentrification of society.

5.   The factor that led most directly to a decline of the central cities was because

        a. the most affluent people moved out.   b. the rural poor moved into the cities.
        c. tax revenues of the cities declined.   d. the welfare costs of the city increased.

        Indicate whether the following events increased [+] or decreased [-] as a result of exurban
        migration?

6.   [   ] industrial development in the central city.
7.   [   ] job opportunities in the central city.
8.   [   ] social unrest and crime rates within the central city.
9.   [   ] accessibility of jobs outside the central city to its residents.
10.   [   ] level of education offered to and achieved by central city residents.
11.   [   ] types and efficiency of local government services.
12.   [   ] property tax base of the central city.
13.   [   ] unemployment rate of central city residents.
14.   [   ] urban decay
15.   [   ] gentrification
16.   [   ] urban sprawl

        Use the terms MORE or LESS to complete the below statement.

        More urban sprawl leads to 17. _____ exurban migration which leads to 18. _____
urban decay which leads to 19. _____ urban sprawl.

        The Highway Trust Fund

20.   benefits both central city and suburb residents. (True or False)
21.   brings people closer together. (True or False)
22.   promotes the establishment of parks and other natural areas. (True or False)
23.   tends to disrupt existing communities. (True or False)

24.   One effective way to stem urban blight is to

        a. provide mechanisms that give residents ownership in their home and community.
        b. give residents large sums of money to purchase goods and services.
        c. have volunteers come in and fix things up for residents.
        d. use government funds to build large multiple housing units.

25.   List one program that gives central city residents ownership of their homes and community.

        _____

ANSWER KEYS

Chapter 1

ANSWERS TO STUDY GUIDE QUESTIONS

1. society fails to care for the environment and sustain it, population increases beyond food growing capacity, disparity between haves and have nots widens; 2. environmentalists; 3. +, -, +, -, -, +, -; 4. environmental crises, public education; 5. murky and irritating, fowled with raw sewage, decline or extinction; 6. the possessions to fuel American lifestyle; 7. pesticide (DDT) use and high bird mortality; 8. wildlife are the indicators of environmental conditions; 9. Environmental Defense Fund, Greenpeace, Natural Resource Defense Council, Zero Population Growth; 10. Environmental Protection Agency, environmental laws, species saved from extinction, pollution abatement; 11. law, engineering, economics, science; 12. a manufacturing plant or feed lot, fertilizer run-off from urban homes and agricultural fields; 13. environment vs jobs; 14. Wise-use; 15. population growth, atmospheric changes, soil degradation, loss of biodiversity; 16. 5.7 billion, 10 billion; 17. American's consumption of the world's energy and natural resources to produce more possessions is having negative global environmental effects; 18. erosion, depleted water supplies, desertification, urban development, salinization; 19. global warming; 20. 300, 360; 21. CO_2 traps heat escaping from the earth raising the temperature of the earth's atmosphere; 22. habitat conversion, pollution, commercial exploitation; 23. total diversity of living things that inhabit the earth; 24. habitat alteration, exploitation, pollution; 25. genetic derivatives of all agricultural plants and animals, derivatives of medicinal drugs and chemicals, maintain the stability of natural ecosystems; 26. inability to meet the basic human needs, continued loss of earth's resources that support human life; 27. population growth will exceed food producing capabilities; 28. Malthus did not envision new technologies to increase food production; 29. environmental stresses (e.g., human population growth) are growing faster, reaching limits of technological solutions; 30. 9; 31. All -; 32. Optomists concerning human capacity to devise technological solutions for any problem; 33. E, C, C, E, E, C; 34. Continued indefinitely without depleting material or energy resources required to keep it running; 35. Is not; 36. 8k; 37. The 8k represents the surplus. Leave the 2k alone to produce another surplus; 38. continues generation after generation, does not exceed sustainable yields, does not pollute beyond nature's capacity to absorb; 39. Meets the needs of the present without compromising the ability of future generations to meet their needs; 40. protect from erosion, conserve and protect, stabalize, reduce, understand and improve relations, provide opportunities for all, apply to recycling and pollution abatement; 41. anyone who cares for creation, is interested in human sustainability, makes efforts to achieve sustainability.

ANSWERS TO SELF-TEST

1. 4, 1, 2, 3, 5 (may be debate on where the chain starts); 2. Population had no where else to go, had no import capabilities, land mass small and isolated; 3. b; 4. b; 5. d; 6. -, +, +, -; 7. Population growth; 8. environment vs. jobs; 9. 4, 2, 3, 1, 4, 3; 10. Population growth will exceed food producing capabilities; 11. If you check the first four and no others, you are a cornucopian. If you check the last four and no other, you are an environmentalist. If you checked a blend of the first and second group of four - you are confused! 12. d; 13. -; 14. -; 15. +; 16. +; 17. +; 18. -; 18. 19. -; 20. + (depending on your opinion of the American public).

Term	Definition	Example	Term	Definition	Example
biodiversity	d	d	habitat alteration	c	c
cornucopian	e	e	sustainable yields	g	j

VOCABULARY DRILL ANSWERS

environmental vocations	j	i	sustainable society	h	g
environmental movement	b	b	sustainable development	i	h
environmentalist	a	a	sustainable	f	f

Chapter 2

ANSWERS TO STUDY GUIDE QUESTIONS

1. plants, animals, microbe; 2. B, B, A, B, A, A; 3. reproduce, viable; 4. population; 5. plants, animals, microbes, each other, environment; 6. perpetuated; 7. deciduous forest, grassland, deserts, coniferous forest, tundra, tropical rain forest; 8. communities; 9. grassland/forest; 10. deciduous forest; 11. grassland; 12. desert; 13. climate; 14. true; 15. flow of rivers, migration of animals; 16. biosphere; 17. biotic, abiotic; 18. producers, consumers, decomposers/detritus feeders; 19. carry on photosynthesis; 20. light, carbon dioxide, water, glucose; 21. air, rocks, or water; 22. organic; 23. organic, inorganic; 24. light; 25. autotrophs, green plants; 26. heterotrophs, animals and fungi; 27. false, Indian pipe; 28. humans, fish, worms; 29. herbivores; 30. deer, mice; 31. carnivores, omnivores; 32. PR, PR, PY, PR, PY, PY; 33. tapeworm, tick, any living organism; 34. detritus; 35. earthworms, clams, crayfish; 36. fungi, bacteria; 37. grass, grasshoppers, frogs, snakes, a, b, c, d, e; 38. First trophic level: a, f, g, o, Second trophic level: b, h, I, k, n, Third trophic level: c, d, e, j, l, m, Example of food web relationship would be to link all the Second trophic level organisms with all the First trophic level organisms; 39. a, d, d, c, b, c, b, d, d, a, d; 40. total dry weight of all organisms occupying a trophic level; 41. less, less; 42. illustrations should be a pyramid; 43. relationship between bees and flowers; 44. bees and flowers require a close union to accomplish life functions; 45. habitat, niche; 46. birds feeding in different parts of a tree, day versus night-time feeders on same food source; 47. wind, temperature, precipitation; 48. see Fig. 2-19; 49. stressed; 50. death; 51. true; 52. one; 53. light, temperature, water; 54. temperature, precipitation; 55. water; 56. temperature; 57. Alpine tundra; 58. microclimate; 59. a; 60. ocean, desert, mountain range; 61. roadways, dams, cities; 62. c; 63. agriculture; 64. producing abundant food, creating reservoirs and distributing water, overcoming predation and disease, constructing our own habitats, overcoming competition with other species; 65. true

---

ANSWERS TO SELF-TEST

1. plants; 2. animals; 3. microbes; 4. each other; 5. environment; 6. abiotics; 7. plants; 8. animals; 9. b; 10. c; 11. b; 12. 1; 13. 5; 14. 2; 15. 1; 16. 2; 17. 1; 18. all; 19. all; 20. d; 21. energy loss from one level to the next decreases potential for biomass production at higher levels; 22. c; 23. d; 24. a; 25. b; 26. a; 27. a; 28. -; 29. -; 30. +; 31. +; 32. +; 33. + 34. -; 35. -.

VOCABULARY DRILL ANSWERS					
Term	**Definition**	**Example (s)**	**Term**	**Definition**	**Example (s)**
abiotic	b	b	heterotroph	m	c,m,s,t,u,v,w
abiotic factors	kk	b	host	aa	a,c,g,k,l,m,s,t,u ,v,w
autotroph	l	l	inorganic molecule	g	g
biomass pyramid	ee	x	limiting factors	oo	b

biomass	i	i	limits of tolerance	uu	aa
biome	tt	ee	microclimate	rr	dd
biosphere	j	j	mutualism	ff	y
biota	a	a	niche	jj	z
biotic community	a	a	omnivore	w	v
biotic factors	ss	a	optimum	ll	aa
biotic structure	k	k	organic molecule	p	p
carnivore	v	u	parasites	z	w
chlorophyll	o	o	parasitism	hh	w
climate	qq	cc	photosynthesis	n	n
consumers	m	m	population	d	d
decomposer	s	s	predator	x	u
detritus	r	r	prey	y	c,t,v
detritus feeders	vv	s	primary consumer	t	t
ecologist	g	g	producers	l	l
ecology	f	f	range of tolerance	uu	aa
ecosystem	e	e	secondary consumer	u	u
ecotone	h	e,h	species	c	c
food chain	bb	x	symbiotic	gg	y
food web	cc	x	synergistic effect	pp	bb
habitat	ii	e	trophic levels	dd	x
herbivore	t	t	zone of stress	mm	aa

Chapter 3

ANSWERS TO STUDY GUIDE QUESTIONS

1. gases, liquids, solids; 2. life; 3. 92; 4. 17; 5. +, +, -, +; 6. true, false; 7. E, C, C, E, C, E, E, C, C, E; 8. (carbon dioxide, water), (water, water), (oxygen gas, air), (nitrogen gas, air), (phosphate ion, dissolved in water or in rock or soil minerals); 9. 2 and 5, 6, 6, 6, 4 and 5, 2 and 5; 10. false; 11. space, mass; 12. solids, liquids, gas; 13. ability to move matter; 14. kinetic, potential, chemical; 15. -, -, +, +, -, -, +, +, +; 16. kinetic; 17. potential; 18. chemical; 19. calorie; 20. a, a, b, b; 21. L, H, H, L, H, H; 22. photosynthesis; 23. chlorophyll; 24. 2-5%; 25. glucose; 26. raw materials for production of other organic molecules, plant cell respiration, stored for future use; 27. glucose; 28. reverse; 29. P, B, R,

P, B, P, B, B; 30. all are marked; 31. malnutrition; 32. calorie, nutrients; 33. 60% used to produce energy, body growth and maintenance, undigested residue passes out as wastes; 34. false; 35. termites, cows; 36. first; 37. recycled; 38. carbon dioxide, air; 39. photosynthesis; 40. carbon dioxide; 41. fossil fuels; 42. see Fig. 3-14; 43. minerals; 44. soil, water; 45. organic; 46. urine and other wastes; 47. cutting down rain forests, phosphate fertilizer runoff from agricultural lands; 48. see Fig. 3-15; 49. 78; 50. true; 51. ammonium, nitrates; 52. fixation; 53. legume; 54. ammonium, nitrates; 55. nitrogen gas; 56. see Fig. 3-16; 57. true; 58. ammonium, nitrate; 59. second; 60. nonpolluting, nondepleting; 61. third; 62. decreases; 63. greater, greater; 64. greater, greater; 65. 3, 1, 2, 2, 1, 3, 1; 66. standing biomass represents the harvestable surplus that provides a sustainable harvest; 67. 1, 2, 3, 1, 2, 3, 3, 1, 1, 1.

ANSWERS TO SELF-TEST

1. B, Ca, C, Cl, Cu, H, I, Fe, Mg, Mn, Mo, N, O, P, K, Na, S, Zn; 2. death; 3. the element might be stored in body fat or organs; 4. the element may be discharged with other wastes; 5. c; 6. a; 7. carbon dioxide; 8. sunlight; 9. glucose; 10. glucose; 11. a; 12. b; 13. a; 14. b; 15. a; 16. b; 17. the same number of C, H, and O atoms that go into the reaction come out at the end of the reaction; 18. b; 19. b; 20. a; 21. d; 22. c; 23. c; 24. d; 25. a; 26. b; 27. a; 28. N; 29. C; 30. P; 31. c; 32. energy loss from one level to the next decreases potential for biomass production at higher levels; 33. because all nonrenewable resources are available in fixed amounts and recycling converts fixed to infinite

VOCABULARY DRILL ANSWERS					
Term	**Definition**	**Example (s)**	**Term**	**Definition**	**Example (s)**
anaerobic	x	t	malnutrition	u	q
atoms	b	b	matter	a	a
cellulose	v	r	mineral	f	f
chemical energy	n	l	mixture	e	e
compound	d	d	molecule	c	c, d
elements	b	b	natural organics	i	d
energy	k	i	organic molecule	g	d
entropy	q	o	organic phosphate	y	u
fermentation	w	s	overgrazing	z	v
First Law of Thermodynamics	o	m	oxidation	t	i, j, p, s, w
inorganic molecule	h	g	potential energy	m	h, k, l, r, s
kinetic energy	l	i, j	Second Law of Thermodynamics	p, r	n
Law of Conservation of Matter	aa	w	synthetic organics	j	l

Chapter 4

ANSWERS TO STUDY GUIDE QUESTIONS

1. first herbivores would die out followed by carnivores and scavengers; 2. balance sustains natural AND human ecosystems; 3. population; 4. stable; 5. death; 6. ER, BP, ER, BP, ER, BP; 7. increase; 8. higher; 9. equal to; 10. have massive numbers of offspring, low reproductive rate with long-term care of young; 11. true; 12. true; 13. increase; 14. species become endangered when they reach critical numbers; 15. figure shows predictable rises and falls in populations of wolves and moose; 16. increase, decrease; 17. all should be checked; 18. true, humans; 19. all +; 20. rabbit population exploded and overgrazed grassland, wiped out American chestnut, overly effective predators, endemic species displacement; 21. ring-necked pheasant; 22. lack of competition with endemic species, no predators; 23. true; 24. live to maturity, die, migrate; 25. all -; 26. true; 27. specific abiotic conditions, species adaptations for obtaining nutrients, host-specific plant invaders; 28. +,-,-,-,-,-; 29. J; 30. S; 31. J; 32. J; 33. true; 34. primary, secondary, aquatic; 35. 4, 1, 3, 2; 36. provide first layer of humus; 37. 4, 1, 3, 2; 38. temperature too hot; 39. terrestrial; 40. provides populations for invasions of new habitats; 41. grasslands, redwoods, pine forests; 42. +,-,+,+,-,-; 43. For sustainability, biodiversity is maintained; 44. true, true; 45. J; 46. introducing species from one ecosystem to another, eliminating natural predators, losing biodiversity, altering abiotic factors, misunderstanding the role of fire.

ANSWERS TO SELF-TEST

1. ER; 2. ER; 3. BP; 4. BP; 5. BP; 6. ER; 7. S; 8. J; 9. J; 10. J; 11. J; 12. critical number; 13. 4, 1, 3, 2; 14. 1, 3, 2, 4; 15. a suitable soil base to start with; 16. c; 17. c; 18. c; 19. b; 20. d; 21. -; 22. +; 23. +; 24. -; 25. -.

VOCABULARY DRILL ANSWERS					
Term	**Definition**	**Example (s)**	**Term**	**Definition**	**Example (s)**
aquatic succession	t	r	host specific	n	m
balanced herbivory	o	n	monoculture	m	l
biotic potential	a	a	natural enemies	j	i, m
carrying capacity	l	k	population explosion	e	d
climax ecosystem	q	p	population density	h	g
critical number	i	h	primary succession	r	o
ecological restoration	v	t	recruitment	b	b
ecological succession	p	o, q, r	replacement level	g	f
environmental resistance	f	e	reproductive strategies	c	c
exponential increase	d	d	secondary succession	s	q, r
fire climax	u	s	territoriality	k	j

Chapter 5

ANSWERS TO STUDY GUIDE QUESTIONS

1. selective breeding, natural selection; 2. false; 3. false; 4. repetitive; 5. c; 6. same; 7. true; 8. survival, reproduction; 9. increase; 10. all +; 11. see Fig. 5-4; 12. the way an organism looks, athletic ability, gentle or aggressive, sensitivity to allergens; 13. DNA; 14. genetic; 15. see Fig. 5-5; 16. two, one; 17. 2, 1, 1, 2; 18. fertilization; 19. genes or alleles; 20. males have only one X chromosome that may or may not carry the allele - can never be safe in a heterozygous condition; 21. twins are the result of sexual reproduction whereas clones are the result of asexual reproduction; 22. change in nucleotide sequence; 23. true; 24. slow; 25. 4, 2, 3, 1; 26. all true; 27. long, short; 28. finches; 29. all; 30. d;. 31. giant tortoise, termites, bison, kangaroo; 32. nature can only modify pre-existing species; 33. true; 34. true; 35. cataclysmic event, massive species extinctions, development of new species, period of stability - process can repeat with another cataclysmic event; 36. balanced; 37. adapt, move, die; 38. d; 39. some, enough; 40. all plus; 41. cockroaches (all +), elephants (all -); 42. cockroaches; 43. cockroaches; 44. thin, movement; 45. tectonic; 46. midoceanic ridges, volcanoes, earthquakes, mountains; 47. the original continental land mass that separated into the major continents as they are presently known; 48. movement of continents to different positions on the globe alters climates AND direction and flow of ocean currents, uplifting of mountains alters movement of air currents; 49. 3 P.M. on December 31, last 2 minutes of the year, last 2 seconds of the year; 50. chimpanzee; 51. walking erect thus freeing hands for tool manipulation, greater intellect, greater vocalization and communication repertoire; 52. small gene pool with little variability to fall back on, no original genetic stocks to go back to, no source of genes to transfer, one-quarter of all prescription drugs have active ingredients developed from wild plants.

ANSWERS TO SELF-TEST

1. G; 2. T; 3. G; 4. check; 5. blank; 6. check; 7. b; 8. a; 9. b; 10. LESS; 11. LESS; 12. Not all organisms that survive have the physical, behavioral, and other adaptations necessary to reproduce; 13. b; 14. d; 15. the terrestrial herbivore niche was unoccupied; 16 - 19. the **greatest** advantage goats had over giant tortoises was probably their ability to obtain food (i.e., goats are more effective herbivores), goats are also faster on land than tortoises which gave them an additional advantage of escaping from predators; 20 - 25. all checked; 26. check; 27. check; 28. blank; 29. blank; 30. -; 31. -; 32. +; 33. +; 34. human have the ecological criteria for survival but have introduced conditions that make the ecosystem nonsustainable for them and all other species; 35. movement of the earth's crust (i.e., tectonic plates).

VOCABULARY DRILL ANSWERS					
Term	**Definition**	**Example (s)**	**Term**	**Definition**	**Example (s)**
adaptation	k	i	mutation	o	l
alleles	n	k	natural selection	g	e
biological evolution	d	d	selective breeding	e	c
differential reproduction	c	c	selective pressures	f	e
DNA	l	j	speciation	i	g
gene	m	k	tectonic plates	j	h
gene pool	b	b	trait	h	f
genetic variation	a	a			

Chapter 6

ANSWERS TO STUDY GUIDE QUESTIONS

1. 1, since the beginning of human history; 2, 100; 3, 30; 4, 15; 5, 12; 6, 11; 7, 11; 8, 11; 9, 13; 10, 13; 2. low recruitment; 3. cause of diseases recognized, improvements in nutrition, discovery of penecillin, improvements in medicine; 4. 12; 5. A, B, C; 6. 21, 80; 7. 79, 20; 8. malnutrition, illiteracy, disease, squalid surroundings, high infant mortality, low life expectancy; 9. Kenya, Italy; 10. decreased, increased; 11. b, a, b, a, a, b, a, b, b, b, a, a, a,; 12. environmental; 13. stabalize population, decrease consumptive lifestyles, increase environmental regard; 14. subdivide farms, open new land for agriculture, move to cities, engange in illicit activities, emigrate to other countries; 15. all +; 16. new suitable land for agriculture left to open; 17. all +; 18. drugs and poaching; 19. developed; 20. emigration; 21. more; 22. false; 23. true; 24. false; 25. 4.5, 25%; 26. developed countries mine the resources of developing countries and leave their wastes in developing countries; 27. answer depends on your point of view; 28. population growth and demand for goods and services; 29. -, 0, +; 30. =2; 31. =2, >2, <2; 32. column; 33. pyramid; 34. births and deaths; 35. less; 36. more; 37. less; 38. less; 39. more; 40. less; 41. more; 42. more; 43. 50 to 60; 44. inverse relationship between economic development and fertility rates; 45. increase; 46. 70; 47. see Table 6-2; 48. long; 49. births and deaths; 50. four; 51. +, +, +; +, -, +; -, -, 0; 0, 0, 0; 52. IV; 53. III.

ANSWERS TO SELF-TEST

1. draw a J-shaped curve; 2. L; 3. H; 4. L; 5. H; 6. 25, 80; 7. 3; 8. 75, 20; 9. 0.26; 10. 1.6; 11. 0.8; 12. considerable change in quality of life; 13. very small change in resource availabilty; 14. 2.4, 30; 15. 0.5, 137; 16. L; 17. B; 18. B; 19. L; 20. make projections on future population trends; 21. IV; 22. c; 23. d; 24. a; 25. population momentum.

VOCABULARY DRILL ANSWERS					
Term	**Definition**	**Example (s)**	**Term**	**Definition**	**Example (s)**
absolute poverty	l	a	doubling time	i	e
age structure	g	h	environmental regard	j	f
crude birth rate	e	b	natural increase	h	g
crude death rate	f	b	population profile	o	h
demography	b	h	population momentum	c	i
demographic transition	d	c	replacement fertility	a	j
developed countries	m	d	total fertility rate	k	k
developing countries	n	l			

Chapter 7

ANSWERS TO STUDY GUIDE QUESTIONS

1. population growth will exceed food-producing capacity of nations; 2. agricultural technology, declining fertility rates; 3. controlling population, economic development; 4. incomes; living; markets; trade, cooperation, peace; fertility; 5. 50; 6. -, +, +, -, -, -, -; 7. -, +, -, -; 8. all -; 9. growing; 10. unpaid interest is added to principal - shrink deeper into debt; 11. cash crops, austerity measures, exploitation of natural resources; 12. they liquidate capital assets to raise cash for short term needs; 13. electric power plants, transportation - super highways, mechanized agriculture; 14. false; 15. true; 16. A, A, B, B, D, A, B; 17. check first three, last one blank; 18. all checked; 19. +, +, +, +, -, +, +; 20. all checked; 21. credit risks, loans too small, status of women; 22. micro lending; 23. women; 24. true, false, true, false, true; 25. W, G, W, G, G; 26. productivity, education, marriage, children; 27. to create an economic, social, and cultural environment in which all people, regardless of age, race, or gender, can equitably share a state of well-being; 28. 7% of each nation's GNP.

ANSWERS TO SELF-TEST

1. C; 2. A; 3. A; 4. +; 5. -; 6. -; 7. +; 8. +; 9. -; 10. +; 11. +; 12. the people of developing nations were unable to enhance their quality of life by moving to the city and finding jobs which led to high fertility rates and a population explosion; 13. low wages; 14. weak child labor laws; 15. weak environmental laws; 16. exploits the poor by keeping them poor with low wages and social and environmental degradation; 17. +; 18. -; 19. +; 20. -; 21. Grumman; 22. availability of education for women; 23. availability of contraceptives; 24. good health and nutrition for mothers and children; 25. c.

Chapter 8

ANSWERS TO STUDY QUESTIONS

1. -, -, -, -, -, +, +, +; 2. industrial; 3. 3; 4. +, +, +, +, +, -, -, -; 5. -, -, -, -, -, +, +, +; 6. true, true, false; 7. true, true; 8. false, true, true; 9. all true; 10. true, false, true; 11. fertilizer, water, double, triple; 12. green; 13. B, D, D, D, D, D; 14. more, less, more, more; 15. 25%; 16. 70; 17. all +; 18. continue to increase crop yields, grow more food instead of cash crops; 19. loss of cropland through soil erosion, climate changes; 20. North America; 21. b, c, c, c, c; 22. population growth surpassed food production; 23. family, nation, globe; 24. b, c, a, a, c, b; 25. 20; 26. 3, 1, 2; 27. poverty; 28. true; 29. economic, nutritional; 30. more, less; 31. war, poverty; 32. a, 100K, b, 900K; 33. -, +, +, +; 34. disrupts, decreases; 35. aggravate; 36. intense irrigation that led to salinization of soil, overgrazing and deforestation; 37. producing enough food to feed the human population without ruining any part of the supporting system or degrading the environment; 38. 1, 4, 2, 3; 39. b.

ANSWERS TO SELF-TEST

1 through 5. all blank, cannot rely on any of these methods to increase agricultural production in the next 40 years. 6. there is not additional tillable land left to bring into agriculture; 7. the optimum level is already being used, using more will not increase production; 8. ground table water reserves are nearly depleted, no other large supplies of water left; 9. pest resistance capabilities and adverse effects on humans and the environment preclude continued use of pesticides; 10. there are no new laboratory designed "high-yielding" varieties to use; 11. blank; 12. +; 13. +; 14. +; 15. +; 16. blank; 17. population; 18. family; 19. nation; 20. global; 21. surpluses are distributed based on economic priorities rather than the nutritional needs of people, i.e., more money can be made from turning surpluses into cat food than feeding starving people; 22. free food aid undercuts the peasant farmer's ability and willingness to grow crops and free food is not distributed equitably or to those who need it most.

Term	Definition	Example (s)	Term	Definition	Example (s)
VOCABULARY DRILL ANSWERS					
subsistence farming	n	i	sustainable agriculture	o	j
food security	f	e	organic farming	m	h
hunger	j	h	green manures	h	g
malnutrition	l	h	green revolution	g	f
undernutrition	p	h, k	cash crops	b	b
absolute poverty	a	a, h	exports	c	b
food aid	e	d	high-yielding	i	f
famine	d	c	imports	k	b

Chapter 9

ANSWERS TO STUDY QUESTIONS

1. a, a, b; 2. undercuts local production, long-term dependency, displaces farm labor; 3. 1.4 million; 4. erosion, salinization; 5. cutting forests, plowing grasslands, draining wetlands; 6. 90, soil; 7. soil; 8. soil organisms, detritus, mineral particles; 9. detritus; 10. mineral nutrients, water, oxygen; 11. phosphate, potassium, calcium; 12. rocks, weathering; 13. false; 14. detritus; 15. true; 16. all true; 17. -, +, +, -; 18. roots; 18. a; 19. overwatering, compaction; 20. acidity, alkalinity; 21. c; 22. 7; 23. out, in, equally in both directions; 24. good supply of nutrients and nutrient-holding capacity, good infiltration and water-holding capacity, porous structure for aeration, ph = 7, low salt concentration; 25. 40, 40, 20; 26. +, -, -, +, -; 27. more; 28. D, H; 29. b; 30. d; 31. structure; 32. all true; 33. all false; 34. all +; 35. plants get needed nutrients from fungal decomposition of detritus, fungus gets some nutrients from plant roots; 36. true; 37. true; 38. -, -, +, -, -, +, -; 39. wind, water; 40. all true; 41. 3, 1, 2; 42. coarse sand and stones; 43. clay and humus; 44. desert; 45. false; 46. overcultivation, overgrazing, deforestation; 47. weed; 48. wind, water; 49. false; 50. 11 to 346; 51. -, +, -, +, +, -, +; 52. false; 53. +, -, +, +; 54. 62; 55. all true; 56. too many cattle on grassland because grazing lease cost so low; 57. all true; 58. a; 59. less, less, less, more; 60. 1; 61. 23; 62. +, -, +, +, -; 63. irrigation; 64. salinization; 65. 30, 2.5, 3.7; 66. Sudan in Africa; 67. 20, salinization; 68. erosion, overgrazing, deforestation, salinization; 69. c, a, b, d; 70. reduce use of chemicals, keep food safe and wholesome, maintain a productive topsoil, keep agriculture economically viable; 71. a, b, c, a; 72. check, blank, check, check, check, check, blank, check, blank; 73. T, A, T, A; 74. traditional; 75. alternative.

ANSWERS TO SELF-TEST

1. see Figure 9-2; 2 to 17. see Table 9-2; 18 to 21. all excellent; 22. the detritus structure is lost which forms the base of the soil ecosystem food chain and one of the primary soil erosion preventatives; 23. without detritus for detritus feeders and decomposers, the transfer of energy and nutrients to other soil biota cannot take place; 24. +; 25. +; 26. +; 27. +; 28. -; 29. -; 30. +; 31. A; 32. -; 33. T; 34. -; 35. T; 36. +; 37. A; 38. +; 39. A.

Term	Definition	Example (s)	Term	Definition	Example (s)
			VOCABULARY DRILL ANSWERS		
compaction	n	g	organic fertilizer	f	c, j
composting	t	k	overcultivation	ee	s
deforestation	aa	o	overgrazing	ff	t
desertification	bb	p	pH	o	h
erosion	ii	q	salinization	gg	u
evaporative water	l	d	sediments	hh	v
fertilizer	e	c	soil structure	u	l
gully erosion	cc	q	soil texture	p	i
humus	s	j	soil aeration	m	f
infiltrate	k	d	soil profile	x	l
inorganic fertilizer	g	c	soil fertility	a	a
irrigation	j	e	sheet erosion	jj	q
loam	q	i	splash erosion	kk	q
mineral nutrients	b	b	subsoil	w	l
mineralization	z	n	topsoil	v	l
mycorrhizae	y	m	transpiration	i	d
no-till agriculture	dd	r	water-holding capacity	h	d
nutrient-holding capacity	c	c	weathering	d	c
			workability	r	i

Chapter 10

ANSWERS TO STUDY QUESTIONS

1. false; 2. f; 3. e, 4. d; 5. c; 6. a; 7. b; 8. false; 9. 37, 31; 10. b, a; 11. chemical technology, ecological pest management; 12. chemical technology; 13. ecological pest management; 14. integrated pest management; 15. e, c, d, a; 16. a; 17. a, a, b, c, c, c, a, b, b; 18. pest resistance, pest resurgence, adverse environmental and human health effects; 19. effectiveness; 20. more, new; 21. resistance; 22. more; 23. true; 24. b, a; 25. greater; 26. -, +, +; 27. assumes ecosystem is a static entity in which one species, the pest, can be eliminated; 28. high, reduces, higher; 29. top; 30. fat; 31. 70; 32. false; 33. 3; 34. true, true, true, true, true, false; 35. a, b, a, b; 36. 500, 250, 125, never; 37. d; 38. -, -, +, -, -; 39. cultural, natural

enemies, genetic, natural chemicals; 40. a, b, b, a, a, a,; 41. i, a, h, b, g, c, e, f, d; 42. e, d, c, b, a, c; 43. identification; 44. 90; 45. true; 46. long; 47. more; 48. true; 49. incompatibility; 50. genetic; 51. lethal, repulsive; 52. larva; 53. structural; 54. hooked; 55. false; 56. d; 57. true; 58. screw worm; 59. b, a; 60. isolate, identify, synthesize, use; 61. specific, nontoxic; 62. b; 63. trapping, confusion; 64. true; 65. management; 66. greater; 67. a, b, b, a, c; 68. blank, check, check, blank; 69. social, economic, ecological; 70. long; 71. sometimes; 72. FIFRA; 73. register; 74. testing; 75. inadequate testing, bans on case-by-case basis, pesticide exports, lack of public input; 76. a, c, d, a, a, d; 77. residue, cancer; 78. negligible risk, over caution; 79. 200,000; 80. false; 81. exporting countries inform all potential importing countries of actions they have taken to ban or restrict use of pesticides, conditions of safe use of pesticides are addressed in detail; 82. all checked; 83. prohibition of any pesticide banned in this country, 50% reduction in pesticide use.

ANSWERS TO SELF-TEST

1. any organism that is noxious, destructive, or troublesome; 2 to 6. all e; 7. a; 8. b; 9. d; 10. c; 11. biomagnification; 12. DDT levels in fragile egg shells was high; 13. DDT is known to reduce calcium metabolism; 14. DDT levels in fish-eating birds was higher than other organisms in the food chain; 15. pests develop a genetic resistance to pesticides exemplified by surviving members of the population surviving and reproducing many more pests (resurgence) that are resistance; 16. natural enemies; 17. genetic controls; 18. natural chemicals; 19. cultural controls; 20. genetic control; 21. cultural control; 22. natural chemicals; 23. natural enemies; 24. genetic controls; 25. cultural control; 26. natural chemicals; 27. cultural control; 28. a; 29. b; 30. a.

VOCABULARY DRILL ANSWERS					
Term	**Definition**	**Example (s)**	**Term**	**Definition**	**Example (s)**
biomagnification	k	h	natural chemicals	v	o
broad-spectrum	f	b	nonpersistent	m	j
chemical barriers	w	h	organically grown	z	w
cosmetic spraying	o	l	persistent	g	b
cultural control	u	r	pest	a	a
economic threshold	n	k	pesticide treadmill	l	i
FIFRA	y	u	pesticide	b	b
first-generation	d	d	physical barriers	x	t
genetic control	t	q	resistance	h	e
herbicide	c	c	resurgence	i	f
insect life cycle	q	n	second-generation	e	b
insurance spraying	p	m	secondary outbreaks	j	g
IPM	y	v	sex attractant	r	o
natural enemies	s	p			

Chapter 11

ANSWERS TO STUDY QUESTIONS

1. 71; 2. 97; 3. ice, groundwater, atmosphere; 4. 0.4; 5. water; 6. hydrological; 7. evaporation, transpiration; 8. condensation, precipitation; 9. gas, liquid, solid; 10. hydrogen; 11. kinetic energy, hydrogen bonding; 12. attract; 13. separate; 14. hydrogen bonding; 15. kinetic energy; 16. high; 17. low; 18. distillation; 19. true; 20. evaporation, transpiration; 21. relative; 22. more; 23. high, low; 24. high, low; 25. low; 26. infiltration, runoff; 27. surface; 28. capillary, gravitational; 29. capillary; 30. impervious, ground, water; 31. aquifer; 32. recharge; 33. springs, seeps; 34. surface; 35. c, c, b, b, a, b, b, b, c, c, c, b; 36. soil, porous rock; 37. b, c, a, d, e; 38. changing earth's surface, pollution, over withdrawals; 39. supply, purity; 40. 70, 23, 8; 41. a; 42. ground, 10; 43. 30; 44. 90; 45. all true; 46. all true; 47. all false; 48. true, true, true, false; 49. consumptive, a; 50. 60; 51. drip; 52. 100; 53. less; 54. gray; 55. micro filtration/reverse osmosis, distillation; 56. $3; 57. $0.0l; 58. decreased, increased; 59. -, -, -, -, +, +, +, +, +, -, +, +, +; 60. fertilizer, "cides," bacteria, salt, toxic chemicals, oil and grease, litter; 61. true, true, true, false, false; 62. see Fig. 11-26; 63. are not; 64. natural ecosystem, industry, agriculture, domestic; 65. zoning water use districts.

ANSWERS TO SELF-TEST

1-15. (see Figure 11-3); 16. -; 17. +; 18. +; 19. -; 20. +; 21. +; 22. -; 23. -; 24. -; 25. +; 26. +; 27. +; 28. large quantities and different types of pollutants are part of urban runoff; 29. air pollutants become mixed with water molecules during condensation and precipitation (e.g., acid rain); 30. depletion of groundwater and pollutants in soil that are carried into ground water.

VOCABULARY DRILL ANSWERS					
Term	Definition	Example (s)	Term	Definition	Example (s)
aquifer	aa	n, q	polluted water	k	h
capillary water	p	m	precipitation	t	n, o
channelization	jj	aa	rain shadow	m	j
condensation	e	e	recharge area	z	s
consumptive water use	cc	t	relative humidity	i	f
desalinization	ll	cc	saltwater	j	g
drip irrigation	hh	y	saltwater intrusion	ff	w
evaporation	c	c, n	sinkhole	dd	u
evapotranspiration	s	n	spring	bb	n
fresh water	f	b	Storm water management	nn	dd
gravitational water	w	n, r	surface waters	o	l
gray water	ii	z	surface runoff	v	n, p

VOCABULARY DRILL ANSWERS					
groundwater	v	n, q	transpiration	mm	n
humidity	h	c	water purification	l	i
hydrological cycle	r	n	water cycle	r	n
infiltration	q	n	water table	y	n, q
land subsidence	ee	v	water vapor	g	c
nonconsumptive water use	gg	x	water quality	b	b
percolation	x	n	water quantity	a	a
physical states of water	d	d	watershed	n	k
			xeroscaping	kk	bb

Chapter 12

ANSWERS TO STUDY QUESTIONS

1. pollution; 2. aesthetic, biological, human health, local to global; 3. identifying material or materials causing the pollution problem, identifying the source(s), develop and implement control strategies; 4. f, g, a, b, c, d, e; 5. population, use; 6. c; 7. more, less; less, more; less, more; more, less; less, more; more, less; more, less; 8. benthic, phtoplankton; 9. euphotic; 10. 90; 11. surface; 12. bottom; 13. nutrient; 14. benthic plants; 15. +, +, -; 16. poor; 17. watersheds, wetlands; 18. b; 19. +, -, +, +, -, -, -, -, +, +, +, -, -, -; 20. poor; 21. false, false, true; 22. increases; 23. benthic plants; 24. decomposers, oxygen; 25. anaerobic; 26. -, +, -, -, +, +, +, +, -, -, -, +, +, +; 27. phytoplankton, fish; 28. SAV; 29. true; 30. sewage treatment plants, poor farming practices, urban runoff; 31. nutrient; 32. soil; 33. croplands, overgrazed rangelands, deforested areas, construction sites, surface mining, gully erosion; 34. true; 35. a, a, b, b, b; 36. sediment; 37. 6; 38. all +; 39. negative; 40. all -; 41. 50; 42. red tides, lose of coral reefs; 43. attack the symptoms, get at the root cause; 44. (chemical treatments, phytoplankton resistance), (aeration, expense), (harvesting algae, expense), (dredging, cost and no where to put material); 45. true; 46. agriculture/forestry, urban/suburban runoff, sewage effluents; 47. banning phosphate detergents and advanced sewage treatment, controlling agricultural and urban runoff, controlling sediments from construction and mining sites, controlling stream erosion, protecting wetlands, controlling air pollution.

ANSWERS TO SELF-TEST

1. O; 2. E; 3. E, 4. O; 5. O; 6. O; 7. E; 8. E; 9. O; 10. O; 11. E; 12. E; 13. O; 14. O; 15. croplands; 16. overgrazed rangelands; 17. deforested areas; 18. construction sites; and surface mining, gully erosion; 19. fertilizer from cropland; 20. fertilizer from lawns and gardens; 21. animal wastes from feedlots; 22. pet wastes; and human excrement, phosphate detergents, acid precipitation; 23. decreases light penetration to benthic plants; 24. increases phytoplankton growth; 25. decreases oxygen concentration killing animals; 26. b; 27. c; 28. d; 29. b; 30. -; 31. -; 32. -.

Term	Definition	Example (s)	Term	Definition	Example (s)
benthic plants	g	a, p	phytoplankton	f	j, n
BOD	l	b	point-source	r	k
cultural eutrophication	m	c	pollutant	b	l, n
emergent vegetation	i	a, p	pollution	a	m
euphotic zone	j	d	red tides	q	n
eutrophication	d	e	sediment trap	t	o
non-point source	s	f	SAV	h	a, p
nonbiodegradable	c	g	tidal wetlands	o	q, s
nontidal wetlands	p	h	turbid	e	r
oligotrophic	k	i	wetlands	n	h, q, s

Table title: VOCABULARY DRILL ANSWERS

Chapter 13

ANSWERS TO STUDY QUESTIONS

1. diseases; 2. pathogens; 3. see Table 13-1; 4. drinking, eating food, body contact; 5. is; 6. disinfection of public water supplies, improving personal hygiene and sanitation, sanitary collection and treatment of sewage wastes; 7. outdoor, stream; 8. Pasteur; 9. storm; 10. western; 11. treating, waste, storm; 12. false; 13. sewage; 14. 99.9, 0.1; 15. (debris and grit, rags and plastic bags), (particulate organic material, food wastes), (colloidal or organic material, feces), (dissolved inorganic materials, nitrogen); 16. nutrients, sludge; 17. b, a, a, b, c, c, c, a, a, d, c, b, c; 18. bar; 19. grit, decreased, settles; 20. incinerated; 21. land fills; 22. primary, slowly, settles out; 23. 30 to 50; 24. sludge; 25. biological; 26. aerated; 27. organic; 28. clarifier, aeration; 29. 90 to 95; 30. nitrogen, phosphate; 31. +, -, -; 32. ozone, ultraviolet radiation; 33. 98, 2; 34. anaerobic, composting, pasteurization, lime stabilization; 35. digester, methane, treated; 36. humus; 37. pellets; 38. calcium hydroxide; 39. -, -, -, +, -, -; 40. using effluents for irrigation, reconstructing wetlands, artificial wetlands, greenhouse wetlands, overland flow systems; 41. septic, drain; 42. land, percolation; 43. contamination with industrial wastes, public opinion; 44. false.

ANSWERS TO SELF-TEST

1. disease; 2. eutrophication; 3. a; 4. 0.1%; 5. d; 6. a; 7. b; 8. b; 9. d; 10. c; 11. b; 12. a; 13. c; 14. d; 15. c; 16. d; 17. most do not remove dissolved nutrients; 18. a; 19. industrial wastes; 20. public apathy due to lack of education about wastewater.

Term	Definition	Example (s)	Term	Definition	Example (s)
activated sludge	l	a	iprimary clarifiers	i	j
bar screen	f	i	primary treatment	h	j
biogas	p	b	raw sludge	j	k
biological treatment	k	n	raw sewage	a	l
BNR	m	c	sanitary sewers	d	m
chlorinated hydrocarbons	n	d	secondary treatment	k	n
co-composting	r	e	settling tank	g	o
composting toilet	t	f	sludge cake	q	p
pasteurized	s	g	sludge digester	o	q
pathogens	b	h	storm drains	c	r
preliminary treatment	e	i	treated sludge	p	s

Chapter 14

ANSWERS TO STUDY QUESTIONS

1. caught on fire, mercury poisoning of residents, accidental release of toxic chemicals on residents, toxic chemicals leaked from an old dump site; 2. d, c, b, a; 3. loss of raw materials, chemical wastes from production process, risks of accidents or spills in manufacturing process, used shampoo rinsed down the drain, containers and packaging thrown into landfill (see also Fig. 12-1); 4. true; 5. heavy metals, nonbiodegradable synthetic organics; 6. lead, mercury, arsenic, cadmium, tin, chromium, zinc, copper; 7. metal-workings, batteries, electronics; 8. enzymes, physiological, neurological; 9. plastics, fibers, rubber, paint, solvents, pesticides, wood; 10. nonbiodegradable; 11. b, a, c; 12. halogenated; 13. chlorine, bromine, fluorine, iodine; 14. chlorinated; 15. plastics, pesticides, solvents, electric insulation, flame retardants; 16. bioaccumulate; 17. false, true, true, false; 18. filters; 19. individual, food chain; 20. higher; 21. million; 22. c, a, b; 23. sewers, waterways; 24. Clean Air Act, 1970; Clean Water Act, 1972; 25. decreased; 26. more; 27. c, a, b, b, a, c; 28. c, a, d, d; 29. true; 30. bioaccumulation, health hazards; 31. false; 32. arrive, stay; 33. unregulated, Love Canal; 34. 150; 35. assure safe water, regulating the handling and disposal of wastes, looking for future solutions; 36. Safe Drinking Water Act, 1974; 37. +, +, +, -, -, -; 38. cost of protection against a hazard that does not exist, no provision for systematic monitoring of private wells; 39. superfund; 40. tax on chemical raw materials; 41. 1.6; 42. 8.6; 43. 9.1; 44. identification of toxic wastes, protection of groundwater near the site, remediation of groundwater if contaminated, cleaning up the site; 45. EPA; 46. +, -, -, -, -; 47. false; 48. all -; 49. false; 50. all true; 51. all +; 52. d, a, b, c; 53. 115; 54. 10; 55. homeowners, farmers; 56. produce < 200 lbs hazardous mat; 57. better product and/or materials management, find nonhazardous substitute for hazardous material, clean and recyle solvents and lubricants, putting environmentalists and industrialists on the same team.

ANSWERS TO SELF-TEST

1. pesticides; 2. explosives; 3. acids; 4. gasoline; 5. g; 6. f; 7. e; 8. d; 9. c; 10. b; 11. a; 12. b; 13. bioaccumulation; 14. biomagnification; 15. c; 16. a; 17. b; 18. b; 19. a; 20. c; 21. c; 22. a; 23. d; 24. d; 25. d; 26 - 30. check; 31. blank; 32. does not produce wastes in the first place; 33. filters existing pollution.

VOCABULARY DRILL ANSWERS					
Term	Definition	Example (s)	Term	Definition	Example (s)
bioaccumulation	c	j	ground water remediation	h	k
biomagnification	f	e	NPL	i	i
bioremediation	j	b	organic chlorides	c, d	g
halogenated hydrocarbons	c	g	pollution avoidance	l	a
HAZMAT	a	f	pollution control	k	c
midnight dumping	g	d	product life cycle	b	h

Chapter 15

ANSWERS TO STUDY QUESTIONS

1. c, a, b, a, c, a, a, c, a; 2. disperse and dilute in atmosphere, oxidation by hydroxyl ions (Fig. 15-2); 3. true; 4. concentration, time; 5. +, -; 6. true; 7. concentration, exposure; 8. amount, space, mechanisms of removal; 9. fire; 10. industrial; 11. photochemical; 12. decreases; 13. rises; 14. cooler, thermal; 15. 4,000; 16. all true; 17. true; 18. Clean Air Act, 1970, 1977, 1990; 19. all true; 20. identifying pollutants, demonstrating cause and effect relationships, determining source, developing and implementing controls; 21. true; 22. suspended particulate matter, volatile organic compounds, carbon monoxide, nitrogen oxides, sulfur oxides, heavy metals, ozone, air toxics; 23. combustion; 24. all true; 25. difficult; 26. chronic, acute, carcinogenic; 27. a; 28. true; 29. lead; 30. b; 31. Poland; 32. false; 33. all true; 34. all +; 35. ozone; 36. false; 37. true; 38. +, +, +, -, -, +; 39. coal, gasoline, refuse; 40. is not; 41. S, P, S, P, P, P, S, P, S, P, S, P; 42. NO_2; 43. industry, fossil fuel combustion; 44. Clean Air Act; 45. false, true; 46. particulates, SO_2, CO, NO, O_3, lead; 47. regulate air pollution so criteria pollutants remain below primary standard level; 48. H, L, L, L; 49. 98; 50. filters, electrostatic precipitators; 51. false; 52. steel mills, power plants, smelters, construction sites; 53. region must submit plans based on reasonably available control strategies (RACT); 54. 75; 55. platinum coated beads oxidize VOC to CO_2 and H_2O, oxide CO to CO_2, and reduces some NO; 56. tighten emission standards, develop cleaner-burning fuel, drive less; 57. coal-burning electric power plants; 58. 1,000; 59. reducing SO_2 emissions by 10 million metric tons; 60. industry, household products; 61. prohibition of some consumer goods, withholding federal highway funds; 62. 600,000; 63. 189; 64. control strategies, substituting nontoxic chemicals, giving industry flexibility in meeting goals, emphasis on calculating public health risks, attention to accidental spills; 65. 20, 7; 66. true; 67. all true; 68. formaldehyde, cleaners, pesticides, aerosol sprays, radon; 69. asbestos, smoking; 70. 125; 71. true; 72. increase fuel efficiency, emission-free vehicles, improving mass transit systems, reduce commuting distances.

1. d; 2. b; 3. a; 4. d; 5. d; 6. P; 7. P; 8. P; 9. P; 10. P; 11. S; 12. S; 13. P; 14. P; 15. S; 16. S; 17. d; 18. d; 19. chronic; 20. teratogenic; 21. carcinogenic; 22. b; 23. a; 24. b; 25. e; 26. a; 27. c; 28. d; 29-32. Figure 15-18.

VOCABULARY DRILL ANSWERS					
Term	Definition	Example (s)	Term	Definition	Example (s)
acute	j	d	point sources	p	l
air pollutants	c	h, i, j	primary standard	n	m
ambient standards	h	n	primary pollutants	l	a
area sources	q	k	secondary pollutants	m	a
carcinogenic	k	g	stratosphere	b	o
catalytic converter	o	c	temperature inversion	g	e
chronic	i	b	threshold level	d	f
industrial smog	e	j	trophosphere	a	o
photochemical smog	f	i			

Chapter 16

ANSWERS TO STUDY QUESTIONS

1. 1000; 2. rain, fog, mist, snow; 3. b, a, c, c; 4. hydrogen; 5. 7; 6. acidity; 7. 10; 8. 10; 9. 100; 10. 5.5; 11. 4 to 3, 2.8; 12. sulfuric, nitric; 13. fuels; 14. n, a, a, n; 15. coal; 16. taller, further; 17. enzymes, hormones, proteins; 18. +, +, -, +, +; 19. soil has a natural buffer; 20. false; 21. decrease; 22. limestone; 23. buffering; 24. all true; 25. rapid; 26. V, T, T; 27. d, d, d, d, d, i, i, i; 28. b, a, c; 29. 41. 30. U.S., Great Britain and western European countries, Korea and China; 31. 50%; 32. d; 33. a, e; 34. a, b, e; 35. e; 36. d, e; 37. c, d; 38. d; 39. c; 40. e; 41. b; 42. a; 43. overwhelming; 44. more; 45. study; 46. addressed acid rain and CFCs specifically; 47. reductions in SO2 levels, methods of implementing reductions, emissions allowances and trading, emissions purchasing, reductions in NO; 48. check, blank, check, check, check; 49. absorbed; 50. heat; 51. greenhouse; 52. CO2 in gas bubbles trapped in glacier ice; 53. planetary albedo, sulfate aerosols; 54. 25; 55. increase; 56. ocean; 57. 3; 58. 18; 59. water vapor, water cycle; methane, animal husbandry; nitrous oxide, chemical fertilizers; CFCs, refrigerants; 60. cloud cover, solar intensity, volcanic activity; 61. 1 to 5; 62. 5; 63. regional climatic changes, rise in sea level; 64. all +; 65. nine out of 14 hottest years, wide-scale recession of glaciers, sea level is rising; 66. true; 67. 60; 68. check, blank, check, check, blank, check, blank; 69. skin; 70. ozone; 71. ozone; 72. false; 73. see Fig. 16-15; 74. +, -, -, +; 75. refrigerants, plastic foams, electronics cleaners, aerosol cans; 76. catalyst; 77. see Fig. 16-15; 78. CFCs; 79. 5; 80. 7; 81. 2; 82. more; 83. false; 84. true; 85. south; 86. 50; 87. false, the rest all true; 88. Queensland, Australia; 89. 3/4 Australians will develop skin cancer; 90. false; 91. false; 92. +, -, +, +; 93. anthropogenic sources of Cl and Br in stratosphere will start to decline; 94. true; 95. an agreement by 68 nations to scale back CFC production 50% by 2000; 96. true; 97. 1995.

ANSWERS TO SELF-TEST

1. a; 2. hydrogen; 3. 10; 4. c; 5. d; 6-9. see study guide question #32; 10. coal; 11. sulfur dioxide;
12. gasoline; 13. nitrogen dioxide; 14. coal; 15. carbon dioxide; 16. cattle; 17. methane gas; 18. refrigerants; 19. CFC's;
20. gasoline; 21. nitrous oxides; 22. refrigerants; 23. CFCs; 24. plastic foams;
25. electronics; 26. any substance that takes H ions out of solution or binds up H ions natural substances such as calcium carbonate in aquatic and terrestrial ecosystems act as buffers to increased H ion concentrations in acid rain; 27. C; 28. B; 29. B; 30. A; 31. C; 32. A; 33. B; 34. A; 35. A; 36. A; 37. A; 38. B; 39. B; 40. B.

VOCABULARY DRILL ANSWERS					
Term	**Definition**	**Example (s)**	**Term**	**Definition**	**Example (s)**
acid precipitation	d	l	chlorine reservoirs	m	m
acid deposition	b	f	chlorine cycle	k	e
acids	n	k	chlorofluorocarbon	j	g
anthropogenic	a	c	greenhouse gases	g	c
artifacts	f	i	ozone shield	i	n
bases	o	j	pH	c	a
buffer	e	h	planetary albedo	h	b
catalyst	l	d			

Chapter 17

ANSWERS TO STUDY QUESTIONS

1. improvement of human welfare, environmental protection; 2. degraded environments have a direct negative effect on human welfare; 3. I, I, B, D, D; 4. c, a, d, a, b, a, c, d, b, c, d, b; 5. b/c, d, b, b/c, a; 6. effectiveness, efficiency, equity; 7. true; 8. false; 9. positive; 10. states and nations with the strictist environmental regulations have the highest rate of job growth and economic reforms; 11. information and knowledge; 12. no financial costs and contribute to economic efficiency and environment; directed toward solving specific environmental problems; require some investment of resources but have clear economic benefits even if environmental benefits not considered; 13. M, R, R; 14. to control the environmental regulatory process on business and industry; 15. cost-benefit; 16. greater; 17. externality; 18. new jobs or better health; lose of jobs or poorer health; 19. true; 20. equipment purchase and loss of old equipment, implementing control strategies; 21. false; 22. early; 23. see Table 17-2; 24. estimating costs of damages that would occur if regulation were not imposed; 25. 100, threshold; 26. greatest; 27. true; 28. high, decrease; 29. a, a, a, b, c, c; 30. true; 31. acid rain, ground water pollution; 32. true; 33. pollutant produced in one state has greatest effect in another; 34. 80, 75, 75; 35. all +; 36. 3.6, 50; 37. c; 38. all true; 39. past; 40. future; 41. c, b, d, a; 42. 3; 43. different; 44. technological ignorance, lack of choice, memory of past hazards, overselling safety, morality, lack of control, fairness; 45. public concern; 46. true; 47. complete risk assessment, regulatory decision; 48. d; 49. a, b, b, d, b.

ANSWERS TO SELF-TEST

1. 3, 1, 4, 2; 2. monetary, political; 3. false; 4. human welfare, environmental protection; 5. costs steadily increase with each % level of reduction and become prohibitive after 75%; 6. between 50 and 70%; 7. costs exceed benefits prior to 100% reduction and costs rise rapidly after 75%; 8. area of greatest distance between lines representing value of benefits and cost of pollution control; 9 - 14. a; 15. a; 16. decrease and level off; 17. about 12 years; 18. c; 19. b; 20. a; 21. c; 22. b; 23. d; 24. a.

VOCABULARY DRILL ANSWERS					
Term	**Definition**	**Example (s)**	**Term**	**Definition**	**Example (s)**
benefit-cost	e	h	market approach	b	e
cost-benefit analysis	d	b	outrage	s	i
cost-effective	f	a	policy life cycle	a	j
dose	l	g	regulatory approach	c	d
dose-response	m	g	risk characterization	o	f
ecological risks	q	n	risk	j	c
exposure	n	g	risk analysis	k	c
externality	g	o	risk perceptions	r	l
hazard	i	c	risk management	t	k
health risks	p	f	shadow pricing	h	m

Chapter 18

ANSWERS TO STUDY QUESTIONS

1. commercial; 2. all true; 3. -, +, +, +, +; 4. d, b; 5. Florida, Texas; 6. Lacey; 7. Endangered Species, 1973; 8. -, +, +, -, -, -; 9. wild turkey, bald eagle; 10. sandhill cranes are surrogate parents for whooping crane chicks; 11. property rights and jobs; 12. 1.4, 3 to 30; 13. 500; 14. agriculture, medicine, aesthetics and recreation, commercial, intrinsic; 15. a, b, a, b, b, a, b, a; 16. 30; 17. a wild species that can become an alternative human food source; 18. treatment for leukemia; 19. rosy periwinkle; 20. controls high blood pressure; 21. Brazilian pit viper; 22. true; 23. ecotourism; 24. increases; 25. decreases; 26. 200; 27. destruction; 28. a, a, a, a, b, c, b; 29. animal, religion; 30. speciation, extinction, accumulation; 31. 100; 32. Costa Rica; 33. 9000; 34. 500; 35. 600; 36. tropical rainforest; 37. commercial fishing down, decline in waterfowl populations, disappearance of songbirds from some regions, overall decline in songbird populations; 38. small area limits population size, human intrusions most severe; 39. extinctions, habitat alterations, human population growth, pollution, exotic introductions, overuse; 40. habitat alterations; 41. habitat alteration or overuse; 42. human population growth; 43. exotics; 44. pollution; 45. extinctions; 46. overuse; 47. overuse; 48. c, a, b; 49. inverse; 50. decrease; 51. c; 52. slowly; 53. quickly; 54. 39; 55. decreases; 56. more; 57. house mouse, Norway rat, wild boar, donkey, horse, nutria, red fox, armadillo, pheasant, starling, house sparrow, honey bee, fire ant; 58. increase; 59. discourage; 60. e; 61. food, furniture, perceived medicinal value, exotic pets; 62. a species whose role is absolutely vital for the survival of many other species in an ecosystem; 63. CITES, Convention on Biological Diversity; 64. CITES, Convention, Convention.

1. a; 2. b; 3. a; 4. b; 5. e; 6. d; 7. c; 8. b; 9. a; 10. animal, religious; 11. number; 12. b; 13. -; 14. -; 15. -; 16. -; 17. -; 18. +; 19. +; 20. +; 21. more, more, less; 22. more, more, less; 23. d; 24. commercial; 25. b; 26. a; 27. c.

VOCABULARY DRILL ANSWERS					
Term	**Definition**	**Example (s)**	**Term**	**Definition**	**Example (s)**
anthropocentric	f	j	fragmentation	o	h
biodiversity	d	a	genetic bank	j	a
biological wealth	d	a	instrumental value	c	o
biota	c	a	intrinsic value	g	p
conversion	n	i	keystone species	r	m
cultivar	i	d	simplification	p	b
ecotourism	k	g	speciation	l	e
endangered species	a	l	threatened species	b	k
exotic species	q	n	vigor	h	c
extinction	m	f			

Chapter 19

ANSWERS TO STUDY QUESTIONS

1. forests and woodlands, tropical rainforests, oceans, coral reefs and mangroves; 2. removal; 3. c; 4. all -; 5. sustainable logging, tree plantations, extractive reserves, enlarging the national heritage, local control of forests; 6. 38; 7. all true; 8. economic development, human population growth; 9. +, +, +, +, -, -, -; 10. international; 11. territorial; 12. 200; 13. 100; 14. close fishery until it has time to restore itself; 15. +, +, +, -, +; 16. Japan, Norway, Iceland; 17. claim whale meat is an integral part of their diet; 18. pollution; 19. C, M, C, M; 20. all B; 21. tundra and desert land, wetlands; 22. maintenance of hydrological cycle, modification of climate, absorption of pollutants, transformation of toxic chemicals, erosion control and soil building, pest management, maintenance of oxygen and nitrogen cycles, carbon storage and maintenance of carbon cycle; 23. erosion control, absorption of pollutants; 24. $100000; 25. false; 26. economic, ecological; 27. economic; 28. depreciation of natural resources; 29. asset; 30. decrease; 31. renewable; 32. false; 33. manage, regulate use; 34. true; 35. +, +, -, +, +; 36. access; 37. private ownership, regulated access; 38. sustained benefits, fairness in access rights, common consent of the regulated; 39. b, a; 40. optimal; 41. exceeded; 42. +, -; 43. false; 44. -, -, +, +; 45. +, -, +, +, +; 46. repair ecosystem damage to normal structure and function; 47. thorough knowledge of ecosystem and species ecology; 48. soil erosion, surface strip-mining, overgrazing, desertification, lake eutrophication; 49. 40; 50. western, Alaska; 51. preserve; 52. National Park Service, U.S. Fish and Wildlife Service, National Forest Service, Bureau of Land Management; 53. Private; 54. true; 55. 5; 56. second; 57. false; 58. cutting trees less frequently, leaving wider buffer zones along streams, leaving dead logs and debris, protecting broad landscapes; 59. dismantling environmental legislation that impedes private property ownership rights; 60. true; 61. true, true, false, false; 62. Nature Conservancy; 63. 5.5 million or more.

ANSWERS TO SELF-TEST

1. absorption of pollutants; 2. erosion control and soil building; 3. pest control; 4. carbon storage; 5. d; 6. preservation; 7. a population level above that which would reduce yield by decreased population and below that which would reduce yield by resource competition; 8. economic; 9. L; 10. L; 11. L; 12. S; 13. c; 14. it is a resource owned by many people in common or no one; 15. decreased, increased; 16. whale watching; 17. the new definition required more (21) consecutive days of water inundation which removed 30 million wetland acres that fit the original definition (7 consecutive days); 18. b; 19. c; 20. a; 21. tragedy of the commons; 22. conservation or preservation; 23. restoration; 24. maximum sustained yield.

Term	Definition	Example (s)	Term	Definition	Example (s)
carrying capacity	l	f	natural resources	d	b
commons	i	b	natural services	c	c
conservation	g	e	preservation	h	e
environmental backlash	n	h	private land trust	o	i
environmental accounting	e	d	renewable resource	f	b
extractive reserves	a	a	restoration ecology	m	g
fishery	b	b	tragedy of the commons	j	b
maximum sustained yield	k	f			

VOCABULARY DRILL ANSWERS

Chapter 20

ANSWERS TO STUDY QUESTIONS

1. increasing population, changing lifestyles, disposable materials, excessive packaging; 2. paper, yard wastes, metals; 3. d; 4. government; 5. burning, landfills; 6. 75, 12, 13; 7. b, a, d, a, b, c; 8. all a; 9. siting; 10. no one wants a landfill near their home or neighborhood; 11. increased cost of waste disposal, long-distance transfer of MSW; 12. residents reduce waste output, recycling occurs; 13. A, A, D, D, D, A, A, A; 14. all true; 15. transportation; 16. +, +, +, -; 17. bottle laws, recycling, composting; 18. disposable, packaging; 19. reusing; 20. returnable bottles, yard (garage) sales, salvation army contributions; 21. 6, 50, 90; 22. all true; 23. returnable; 24. increase, decrease; 25. 10; 26. industry; 27. resale, eliminating junk mail; 28. 75; 29. e, d, c, b, a, f; 30. strong incentive to recycle, recycling not optional, residential recycling is curbside, recycling goals are clear, involves local industries, municipality employs knowledgeable coordinator, 31. 13; 32. true; 33. 30; 34. 89; 35. false; 36. carpet fibers, irrigation drainage tiles; 37. humus; 38. true; 39. E, A, B, C, D; 40. several; 41. waste reduction, waste disposal, recycling and reuse.

1. c; 2. b; 3. d; 4. b; 5. a; 6. primary; 7. d; 8. c; 9. secondary; 10. d; 11. b; 12. b; 13. b; 14. b; 15. d

VOCABULARY DRILL ANSWERS					
Term	Definition	Example (s)	Term	Definition	Example (s)
landfill	a	a	secondary recycling	g	e
MSW	f	d	SEMASS	d	b
primary recycling	e	c	waste-to-energy	b	b
resource recovery	c	b			

Chapter 21

ANSWERS TO STUDY QUESTIONS

1. human; 2. power; 3. steam; 4. b; 5. a; 6. 80; 7. internal combustion, oil, crude; 8. 41; 9. 23; 10. A, A, A, D; 11. secondary; 12. a; 13. A, D, D, D, D; 14. 35; 15. 65, heat, thermal; 16. major source of acid rain, flooding of terrestrial habits from dams; 17. transportation, industry, residential, electrical power generation; 18. 28, 23, 13, 36; 19. a, e, e, c; 20. a; 21. electrical; 22. coal, crude oil, natural gas; 23. photosynthetic; 24. +, -, +; 25. reserves; 26. proven; 27. 10, remaining; 28. 10 million, 90 million; 29. 9 million, 81 million; 30. false; 31. true; 32. I, D, D, I, I; 33. Organization of Petroleum Exporting Countries; 34. suspending; 35. increased; 36. I, I, I, D, I, D; 37. true; 38. decreased; 39. c; 40. D, D, D, I, D, I; 41. 50; 42. false; 43. costs of purchases, supply disruption, resource limitations, 44. 96; 45. provides jobs that produce other goods and services, money leaves U.S. as another form of foreign aid; 46. b; 47. increase; 48. Alaska; 49. 20, 50; 50. natural gas, coal, oil shales and tar sands; 51. b, c, b, b, a, a or b; 52. 4.5%, 25%; 53. natural gas; 54. conservation, development of non fossil fuel energy sources; 55. decrease; 56. conservation; 57. true; 58. 80; 59. increase fuel efficiency in cars, cogeneration, use florescent lights, increase home insulation; 60. nuclear and solar power.

ANSWERS TO SELF-TEST

1. 2, 1, 4, 3, 5; 2. a, a; 3. a, b; 4. a, b; 5. a, b; 6. b, b; 7. a, b; 8. a, b; 9. a, a; 10. e; 11. d; 12. a; 13. b or c; 14. b or c; 15. d; 16. e; 17. e; 18 to 24. not checked; 25. checked; 26. solar; 27. a multitude of political, economic, and cultural reasons; 28. derived from ancient photosynthetic processes; 29. coal; 30. oil; 31. cannot be made again in the short term; 32. cannot be used again once burned; 33. burn coal to produce heat to turn a turbine to produce electricity to produce heat for a home; 34. O; 35. O; 36. +; 37. +; 38. -; 39. +; 40. +; 41. +; 42. conservation.

Chapter 22

ANSWERS TO STUDY QUESTIONS

1. optimism; 2. pessimism; 3. c; 4. c; 5. true; 6. control nuclear reactions so that energy is released gradually as heat; 7. a, b; 8. less; 9. energy; 10. heat; 11. fission; 12. 235; 13. isotopes; 14. b; 15. 235; 16. b; 17. radioactive byproducts,

heat, neutrons; 18. chain; 19. 238; 20. 235; 21. uncontrolled, high; 22. true, true, true, true, false, false; 23. b, c, b, c, a, a; 24. fuel elements, moderator-coolant, movable control rods; 25. all true; 26. meltdown; 27. 1000; 28. C, N, C, N, C, N, C, N, N; 29. radioisotopes; 30. subatomic, radiation; 31. radioactive; 32. radioactive; 33. true; 34. indirect; 35. damage biological tissue, block cell division, death, damage DNA molecules, cancer, birth defects; 36. lower; 37. earth's crust, cosmic rays from outer space; 38. 1; 39. disposal of radioactive waste, accidents; 40. radioactive; 41. half; 42. seconds, thousands; 43. [S] = iodine-131, [L] = plutonium-239; 44. iodine-131, plutonium-239; 45. 10; 46. 240; 47. short, long; 48. all true; 49. 130; 50. military and high-level nuclear wastes; 51. deliberate release of radioactive wastes, 20-year nuclear waste discharge area; 52. 600, 1.5; 53. the public response "not in my back yard"; 54. Yucca Mountain in southwestern Nevada; 55. containment; 56. coolant; 57. human error; 58. decreased; 59. equally; 60. 6; 61. false, true, true, false; 62. c; 63. all true; 64. embrittlement, corrosion; 65. breeder and fusion reactors; 66. BR, BR, BR, BO, BR, FU, BO, FU, FU; 67. a general distrust of the technology, skepticism of management, overall safety of nuclear power plants, nuclear waste disposal problems.

ANSWERS TO SELF-TEST

1. c; 2. a; 3. a; 4. b; 5. c; 6. d; 7. c; 8. c; 9. d; 10. b; 11. d; 12. c; 13. d; 14. a; 15. d

VOCABULARY DRILL ANSWERS					
Term	**Definition**	**Example (s)**	**Term**	**Definition**	**Example (s)**
active safety	o	m	half-life	n	i
background radiation	l	h	indirect products	j	g
breeder reactor	s	j	isotopes	c	f
corrosion	r	a	LOCA	h	o
direct products	i	f	LWRs	u	c
embrittlement	q	a	mass number	e	f
enrichment	f	f	passive safety	p	l
fission products	a	d	radioactive emissions	d	b
fission	a	d	radioactive decay	m	i
fuel rods	g	k	radioactive wastes	j	g
fusion	b	e	radioisotopes	i	i
fusion reactor	t	e	rem	k	n

Chapter 23

ANSWERS TO STUDY QUESTIONS

1. kinetic, thermonuclear; 2. renewable; 3. collection, conversion, storage, cost-effectiveness; 4. heating, electricity,

fuel; 5. B, A, A, A, A, B, A, A, B, B; 6. W, W, B, W, B, B, W, W, S, S; 7. passive; 8. 75, 5, 9. true; 10. 25; 11. ignorance; 12. photovoltaic cells, solar trough collectors; 13. solar; 14. b; 15. missing; 16. additional; 17. accept; 18. accept; 19. current; 20. true, false, false, true; 21. 22, 8; 22. less; 23. steam, turbogenerators; 24. brine, fresh; 25. a, a, b; 26. all true; 27. +, +, +, -, -, -, +; 28. wind, water, biomass, ocean thermal; 29. natural; 30. artificial; 31. dams; 32. hydroelectric; 33. +, -, -, -; 34. 2; 35. mills; 36. turbines; 37. -, +, +, +, +, +, -; 38. fire wood, burning MSW, methane production, alcohol production; 39. +, -, +, -; 40. indirect; 41. true; 42. indirect; 43. both; 44. -, -, +; 45. geothermal, tidal; 46. -, -, -, -, -, -, -; 47. both; 48. +, +, +, -, -, -, -; 49. -, +, -, -, -, -, -.

ANSWERS TO SELF-TEST

1. d; 2. a; 3. a; 4. a; 5. a; 6. a; 7. a; 8. a; 9. c; 10. c; 11. c; 12. c; 13. b; 14. b; 15. a; 16. d; 17. d; 18. d; 19. c; 20. b; 21. b; 22. b.

VOCABULARY DRILL ANSWERS					
Term	Definition	Example (s)	Term	Definition	Example (s)
active systems	b	j	OTEC	l	h
bioconversion	k	a	passive systems	c	i
biomass energy	k	a	power gird	e	f
carbon tax	n	k	PV cell	d	b
electrolysis	g	g	solar trough	f	l
flate-plate collectors	a	d	wind farms	j	m
fuel cells	h	c	wind turbine	i	m
geothermal energy	m	e			

Chapter 24

ANSWERS TO STUDY QUESTIONS

1. car ownership, low home mortages, highways; 2. time; 3. +, +, +, -, -, +, +, -, +, +, +; 4. doubled, no change; 5. inner city → suburbs → exurbs; 6. e, d, c, b, a, d, a; 7. urban decay; 8. all true; 9. +, -, +, +; 10. all true; 11. lower; 12. higher; 13. central city; 14. worse; 15. deteriorating; 16. false; 17. true; 18. -, -, -, +, +, -, -; 19. nonsustainable; 20. all checked; 21. reduced outward sprawl, reduced automobile traffic, mass transit; 22. change zoning laws, provide housing near work and service centers, provide bicycle and pedestrian access; 23. provides funding from highway trust fund to construct modes of access for alternative forms of transportation; 24. b, c, a; 25. buildings, land; 26. true; 27. check, blank, all the rest checked.

ANSWERS TO SELF-TEST

1. b; 2. c; 3. d; 4. b; 5. c; 6. -; 7. -; 8. +; 9. -; 10. -; 11. -; 12. -; 13. +; 14. +; 15. +; 16. +; 17. more; 18. more; 19. more; 20. false; 21. false; 22. false; 23. true; 24. a; 25. empowerment zones or habitat for humanity.

VOCABULARY DRILL ANSWERS					
Term	**Definition**	**Example (s)**	**Term**	**Definition**	**Example (s)**
economic exclusion	g	a	gentrification	d	d
empowerment zones	h	g	urban decay	f	f
eroding tax base	e	e	urban blight	f	f
exurban migration	b	b	urban sprawl	a	c
exurbs	c	f			